GEOCHEMICAL BIOMARKERS

Edited by

T. F. Yen
University of Southern California
Los Angeles, California

and

J. M. Moldowan
Chevron Oil Field Research Co.
Richmond, California

harwood academic publishers
chur • london • paris • new york • melbourne

Harwood Academic Publishers

Post Office Box 197
London WC2E 9PX
England

58, rue Lhomond
75005 Paris
France

Post Office Box 786
Cooper Station
New York, New York 10276
United States of America

Private Bag 8
Camberwell, Victoria 3124
Australia

Library of Congress Cataloging-in-Publication Data

Geochemical biomarkers.
  1. Organic geochemistry. 2. Biological markers.
I. Yen, Teh Fu, 1927–   Moldowan, J. Michael,
1946–
QE516.5.G445   1988   551.9   88-2616
ISBN 3-7186-0483-3

In memory of

Wolfgang Klaus Seifert
1931–1985

for his pioneering work in this field

CONTENTS

PREFACE

There are several ways to assess the paleodepositional environment and diagenetic past of sedimentary rocks. When we concentrate our attention on the organic matter within those rocks, we are engaged in the science of organic geochemistry. This organic matter may be studied on several different levels. Its gross fossil or preserved structure is assessed through micropaleontology, and classical organic geochemistry deals with bulk properties of the organic matter. High resolution chromatographic methods often coupled with mass spectral analysis allow us to separate and analyze the organic components to the extent that they may be studied on the molecular level.

Thus another area of organic geochemistry has opened up in recent years which concentrates on the study of molecular fossils, or as they are sometimes called, geochemical biomarkers. These are complex organic molecules found in the geosphere which have survived or retained enough chemical structural information to be recognizable as having a biochemical origin.

Certain molecular structural features may be directly linked to a given type of biotic input. Acid catalyzed, photochemically or enzymatically induced alterations in biomarker structure generally reflect early diagenetic processes while higher energy, free radical and acid catalyzed, transformations reflect later diagenesis and catagenesis of the organic matter. In fact, the progress of these later reactions can be used to monitor the thermal history of a given sediment.

An economically important aspect of biomarker chemistry is that the same biomarkers and their transformations can be measured in crude oils as in sedimentary rock extracts, including crude oil source rocks. Thus biomarkers can be used to obtain many kinds of information useful in petroleum exploration such as maturity of potential source rocks and relative maturities among oils and rocks, aspects of environment of deposition of an oil source rock, including the types of organisms contributing to the lipids and the strength of anoxicity at the sediment water interface during deposition.

In this volume, the authors address a number of recent advances in the fundamental knowledge and applications of biological markers. These advances cover a wide spectrum of interests within the field ranging from the property exhibited by the natural asphalts and bitumen systems to its formation of different ranks of coals, from the diagenetic fate of biological compounds in natural waters and aerosols to the recognition of maturation effects in ancient sediments, and from the characterization of depositional environments of petroleum source rocks to the classification of precursor organisms by their molecular fossil remnants.

As a point of interest, this volume contains the most extensive collection of

xi

chapters on geochemical markers of the continental basins of China yet to appear in the western literature. These Chinese basins provide a unique laboratory for geochemical research on paleo-lacustrine systems having diverse depositional environments.

Most chapters in this volume were originated initially from a Symposium entitled "Geochemical Biomarkers," sponsored by the Geochemistry Division of the American Chemical Society, at which, the present editors served as co-chairmen.

The authors would like to thank Seon-Hwa (Deseree) Moon, Nancy Yan and Linda Raftree for their editing assistance.

T. F. Yen
University of Southern California
Los Angeles, California

J. M. Moldowan
Chevron Oil Field Research Co.
Richmond, California

BIOLOGICAL MARKERS IN GRAHAMITES AND PYROBITUMENS

J. A. CURIALE
Unocal Science & Technology Division
Unocal Corporation
376 South Valencia
Brea, California 92621

Grahamites and pyrobitumens are solid bitumens
traditionally distinguished on the basis of their
solubilities; pyrobitumens exhibit low solubility in
carbon disulfide. In this paper biomarker analyses
were used to examine grahamites from Oklahoma,
Pennsylvania and Spain, albertites from Utah and
Canada, impsonite from Oklahoma, and ingramite from
Utah. Although both grahamites and pyrobitumens
contain a suite of biomarkers commonly associated
with petroleum, including steranes, diasteranes and
terpanes, relative concentrations of steroidal
hydrocarbons vs terpanes are generally very low in
comparison with crude oils. As with conventional
crude oils, biomarker distributions of the extractable organic matter of solid bitumens are useful for
bitumen-bitumen and oil-bitumen correlations, and
assessment of source type and thermal maturation.
Both 20S/20R- and 14β,17β(H)/14α,17α(H)-
ethylcholestane ratios are useful in assessing solid
bitumen maturation. Results suggest that biomarker
analyses of grahamite and pyrobitumen extracts can
assist the petroleum explorationist in assessing
basinal prospects.

1. INTRODUCTION

Solid bitumens are defined, for the purposes of this
paper, as non-coaly, allochthonous, solid, localized
organic matter found in, or associated with, rocks. These
materials are found throughout the stratigraphic section
in sedimentary basins from around the world. The im-
plications associated with their occurrence have often
puzzled the petroleum explorationist: is the presence of
solid bitumen noteworthy when exploring for oil and gas?
What is the origin of this material? Can knowledge of
the chemical composition of solid bitumens be applied to
the search for petroleum? It is this last question that
this paper addresses. More specifically, I will discuss
the characteristics of certain common biological markers
found in the solvent extractable fraction of two particular kinds of solid bitumens. Our knowledge concerning
applications of biomarker technology to petroleum exploration can be expanded by studying the cyclic hydrocarbons in solid bitumens.

Solid bitumens have been traditionally classified in
a generic fashion, utilizing carbon disulfide solubility,
flame fusibility and hydrogen/carbon atomic ratios as
classification criteria. Thus, "grahamite" is the name
applied to a solid bitumen which is relatively soluble in
carbon disulfide, yet fusible only with difficulty,
whereas "albertite" is applied to those solids which are
relatively insoluble in carbon disulfide, having H/C
ratios greater than 1.0. As a class, solid bitumens
having very low carbon disulfide solubilities are referred to as pyrobitumens, and include albertite, impsonite

and ingramite. A generic classification scheme such as
this, while useful to those interested in the technolo-
gical utility of solid bitumens, is of little use to the
petroleum explorationist. Rather, the explorationist is
concerned with the origin of the material. For this
reason, a genetic classification scheme, based on organic
geochemical criteria, is the classification method of
choice for solid bitumens[4].

As part of a continuing effort to define the origin
of various solid bitumens, the biomarker distributions of
two generic classes of these materials, grahamites and
pyrobitumens, have been examined. Previous work has
suggested that, despite the largely insoluble nature of
many of the samples involved, sufficient material can be
extracted for geochemical analysis.[1,11,13,28] Terpanes
and steroid hydrocarbons, including steranes, diasteranes
and aromatized steroids, were examined.

2. ANALYTICAL METHODS

A total of thirteen samples (Table I) were studied,
including grahamites, albertites, ingramites and impsonites. Sample preparation methods are the same as those
described for rock samples by Curiale;[6] extractions were
accomplished with a ternary solvent, having 70% toluene,
15% methanol and 15% acetone. Biological markers were
analyzed in the total gas oil hydrocarbon fraction (with
prior removal of n-paraffins), rather than as separate
aliphatic and aromatic hydrocarbon fractions, in order to
directly compare steranes and terpanes with aromatized

steroid hydrocarbons. Analyses were conducted on a
Finnigan MAT Triple Stage Quadrupole. Instrument conditions are as in Curiale[6], excepting that analyses for
solid bitumen hydrocarbon fractions were carried out
using the MID mode, monitoring

TABLE I.   Sample Identification

Sample	Location	Age	Generic Name
AK7429	Sardis Mine east, OK	Miss [a]	grahamite
AK7430	Jumbo Mine, OK	Miss	grahamite
AK7431	Sardis Mine west, OK	Miss	grahamite
AK7432	Pumroy Mine, OK	Miss	grahamite
AK7433	South Bald Pit, OK	Miss	grahamite
AK7434	Page Mine, OK	Miss	impsonite
AK7442	Washington County, Penn.	---	grahamite
AK7443	Page Mine, OK	Miss	impsonite
AK7444	Wasatch County, Utah	Eocene	albertite
AK7445	Wasatch County, Utah	Eocene	ingramite
AK7446	New Brunswick, Canada	Devonian	albertite
AK7452	Northern Spain	---	grah/imps
AK7453	Jumbo Mine, OK	Miss	grahamite

a Mississippian

parents and major fragments of the common biomarkers (a
total of 23 ions). Relative distributions of biological
markers discussed herein are based on area integrations
of major fragment peaks. Comparisons of biomarkers are

based on the relationship between integrated areas taken
from several chromatograms, and are uncorrected for
relative fragment intensities for separate compound
types.

A total of fifty-eight cyclic hydrocarbons were
monitored. These include the 13β,17α-diasteranes having
27-29 carbon atoms; five regular steranes of each carbon
number from C_{27} through C_{29} (5α,14α,17α(H),20S&R;
5α,14β,17β(H),20S&R; and 5β,14α,17α(H),20R); major
monoaromatic and triaromatic steroid hydrocarbons; and
tricyclic and pentacyclic terpanes, through the bishomo-
hopane epimers. This 58-compound distribution is histo-
gram-formated, for the purposes of this paper, as shown
in Figure 1: from left to right, diasteranes, steranes,
monoaromatic steroids, triaromatic steroids, tricyclic
terpanes, and pentacyclic terpanes. Specific compound
identifications are presented elsewhere[4].

3. RESULTS

Biological marker concentrations in the grahamites and
pyrobitumens vary with the content of extractable organic
matter (EOM) in each sample. Percentage EOM values, and
two distinctive biomarker ratios are shown in Table II.
The values indicate an EOM range for grahamites and pyrobitumens of 12.4-25.8% and 0.5-7.1%, respectively. Note
that, based on EOM yields, the sample having an unclear
generic designation, AK7452 (Table I), is probably the
pyrobitumen impsonite.

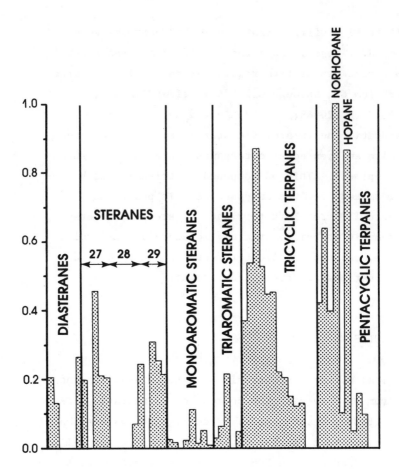

FIGURE 1   Distribution of fifty-eight (58) biological
markers in sample AK7432, an Oklahoma grahamite.

TABLE II   Numerical Data

Sample	EOM(%)	a	b
AK7429	16.0	--	--
AK7430	22.9	0.50	0.35
AK7431	16.9	0.48	0.29
AK7432	25.8	0.55	0.52
AK7433	23.7	0.45	0.37
AK7434	0.5	--	--
AK7442	12.4	0.61	0.30
AK7443	1.2	--	--
AK7444	7.1	--	0.83
AK7445	6.8	--	0.77
AK7446	2.8	0.39	0.19
AK7452	2.7	--	--
AK7453	13.2	0.46	0.34

a. $(5\alpha,14\beta,17\beta,20R\&S/5\alpha,14\beta,17\beta,$
   $20R\&S+5\alpha,14\alpha,17\alpha,20R\&S)$
b. tricyclic terpanes/pentacyclic
   terpanes

Despite the fact that the grahamites and pyrobitumens examined represent a wide range of biomarker
characteristics, certain common features are present
in some samples.   For example, concentrations of pentacyclic terpanes are equal to or greater than those
of tricyclic terpanes in all samples except the Oklahoma
impsonites (containing only diasteranes) and the Utah

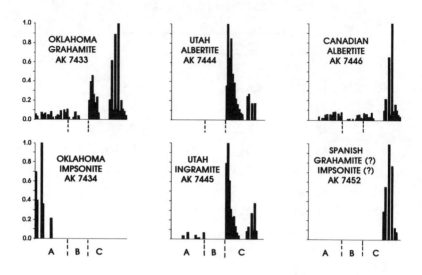

FIGURE 2 Biological marker fingerprints for extractable organic matter from six grahamites and pyrobitumens. Compound order A: steranes. B: Aromatic Steroids. C: Terpanes.

pyrobitumens (AK7444, AK7445) (Table II).
Monoaromatic and triaromatic steroid hydrocarbons are either absent or in extremely low concentration in all grahamites and pyrobitumens examined. In the single sample containing both monoaromatic and triaromatic steroid hydrocarbons (AK7446), the triaromatic compounds dominate.

These characteristic features are shown together in Figure 2, for six solid bitumens. The distributions in Figure 2 appear to be controlled by at least two processes, namely source input and subsequent organic matter alteration. For example, the similarities in the bio-

marker distributions of the two Utah solid bitumens,
albertite and ingramite, suggest a common source, which
is undoubtedly the Green River Formation of the Uinta
Basin.[15] On the other hand, the Oklahoma grahamite and
impsonite distributions in Figure 2 are significantly
different, despite the fact that they have been shown to
be commonly sourced.[7,8] Consequently, differences in
their biomarker distributions must be attributable to
organic matter changes subsequent to the time of sourc-
ing. In the case of the impsonite, these changes are the
result of the fact that the impsonite is located in an
extreme paleothermal regime, which probably resulted in
either cracking and destruction, or the incorporation
into the insoluble fraction, of most steroid and all
terpenoid hydrocarbons[7]. It is interesting that the
distribution of steroidal hydrocarbons in the extractable
organic matter from this pyrobitumen is very similar to
that in the pyrolyzate[9].

4. DISCUSSION

Biological markers have found recent applications in
several aspects of petroleum geochemistry. They are most
commonly used as source organic matter indicators, in an
oil-oil or oil-source rock correlation fashion.[6,24,26]
In addition, their uses as maturation parameters[5] and
indicators of biodegradation[2,23,25] are now commonplace.
The results of the present study of biomarkers in solid

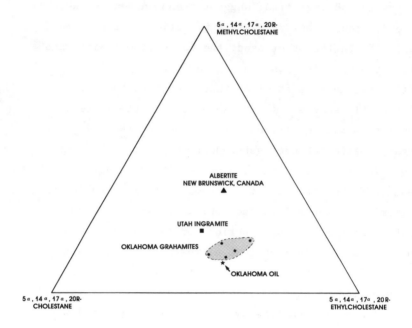

FIGURE 3   Ternary diagram showing the distribution of
5α,14α,17α,20R steranes by carbon number (C_{27}, C_{23}, C_{29}),
for seven grahamites and pyrobitumens.

bitumens show that these common uses in petroleums and
source rock extracts are immediately extendable to the
extractable organic matter content of grahamites and
pyrobitumens. Consequently, biological markers can be
utilized for bitumen-bitumen correlations, estimates of
thermal maturity of solid bitumens, and determination of

the post-expulsion effects of biodegradation on these
materials[9]. Taken together, knowledge of the biological
marker distribution in the EOM of grahamites and pyrobitumens provides a powerful tool for elucidating the
genesis of these solids in sedimentary basins.

4.1 Correlation Applications

Recent work involving sterols in recent sediments has
suggested that the distribution of sterol carbon number
in sediments is indicative of depositional environment[14].
This work has been extended to steranes found in ancient
organic matter, in an effort to deduce source rock
depositional environment, and concurrently provide a
correlation tool (presumably) independent of subsequent
changes due to thermal maturation or other types of
alteration [6,20,21]. Figure 3 shows the sterane carbon
number distributions for the $5\alpha,14\alpha,17\alpha,20R$ steranes, for
five grahamites and two pyrobitumens.

   The separation of the three sample groups suggests
differing source controls for the Canadian albertite,
Utah ingramite and Oklahoma grahamites. Further, the
similar sterane carbon number distributions for the
Oklahoma grahamites implies a common source for these
samples, a conclusion previously reached using bulk
geochemical parameters such as elemental concentrations
and stable isotope ratios[7,8], as well as pyrolysis-GCMS
data[9]. The similarities among the Oklahoma grahamites
based upon $5\alpha,14\alpha,17\alpha,20R$ sterane carbon numbers are
clearly seen in the entire sterane distribution as well.
M/z 217 mass chromatograms for four of these bitumens are

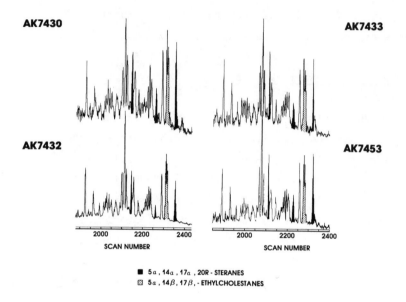

FIGURE 4   M/z 217 mass chromatograms for four Oklahoma grahamites.

shown in Figure 4. Note that the distributions of regular and rearranged steranes are extremely similar from sample to sample, regardless of the fact that solubilities of these solid bitumens range from 13% to 26%. The distributions of terpanes in these four solid bitumens (Fig. 5) are not quite as similar as those of

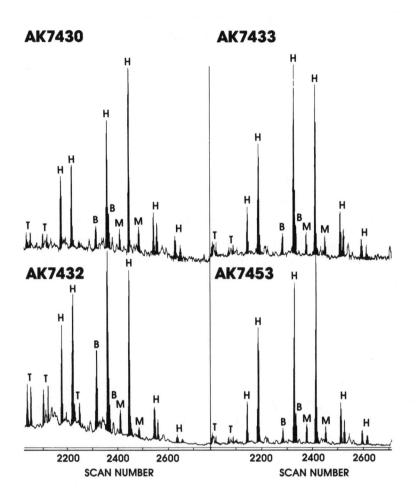

FIGURE 5   M/z 191 mass chromatograms for the same samples shown in Figure 4.

the steranes, although common characteristics are certainly present. One unusual difference among the samples
is the variability in the relative concentration of the
C_{28} and C_{29} tricyclic terpanes (Fig. 5). This variability tends to coincide with solubility trends of the
grahamites: the most soluble samples appear to contain the highest relative amounts of the higher carbon
number tricyclic terpanes (Table II). Another component
which appears to co-vary with tricyclic terpane content,
one which is clearly visible in Figure 5, is the relative
concentration of 28,30-bisnorhopanes. These compounds,
at least two isomers of which are apparent in these
chromatograms[22], are in highest concentration in AK7432,
also the most soluble of the Oklahoma grahamites. Further
examination of this sample, and subsequent examination of
the other Oklahoma grahamite extracts, revealed the
presence of the unusual compound 25,28,30-trisnorhopane.
This compound (Fig. 6) has been noted in several California Monterey oils, and well as extracts from North Sea
rock samples and recent sediments in the Norwegian Sea[12].
Nevertheless, its occurrence is generally regarded as
unusual[6]. In addition to bitumen-bitumen correlation
utilizing extractable biomarkers, it would also be useful
if these compounds could be used for oil-solid bitumen
correlation efforts. That this is feasible is shown by
the close correlation between the Oklahoma grahamites and
an oil produced from similar age rocks nearby. This oil
(Oil #45 of ref. 7) has been shown previously, using bulk
geochemical parameters, to be commonly sourced with these
Oklahoma grahamites[10]. The ability to geochemically

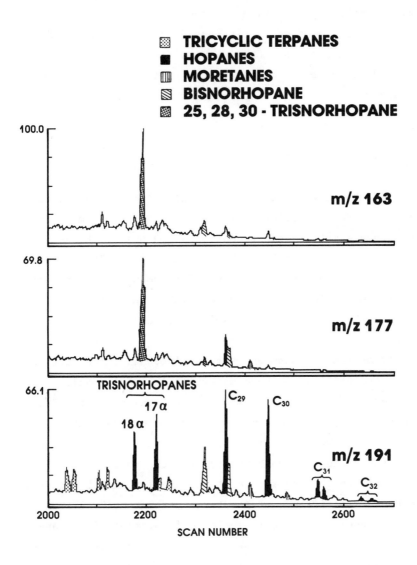

FIGURE 6   M/z 163, 177 and 191 mass chromatograms for sample AK7432, an Oklahoma grahamite exhibiting the highest relative concentration of 28,30-bisnorhopanes.

correlate a solid bitumen with a crude oil using biological marker data provides a specific method for assessing the origin of solid bitumens encountered during the
exploration process. Continuing efforts in this area
include attempts to correlate the pyrobitumens of this
study with crude oils from similar locations.

4.2   Thermal Maturity of Solid Bitumens

One of the earliest uses of biological marker geochemistry, the determination of the so-called thermal maturity
of organic matter, has recently become one of the most
fruitful aspects of biomarker research[5,19]. The nature
of the two major origins of solid bitumens (i.e., altered
petroleum vs immature petroleum) makes the determination
of the thermal maturity of solid bitumens of critical
importance. The maturity question is of particular
interest for pyrobitumens such as impsonite, insomuch as
the insolubility of these materials is traditionally
viewed as the result of extensive thermal alteration[16].
Consequently, the common biomarker thermal maturity
parameters, such as epimer ratios, have been calculated
for the solid bitumen extracts, for the purpose of
determining the maturity of these samples and the genetic
connotation, if any, of the term "pyrobitumen". Although
the biochemical stereoisomeric configuration of steranes
at the 14 and 17 positions is $\alpha\alpha$, empirical observations
have shown that this isomer converts to the $\beta\beta$ configuration with increasing thermal input[17,24]. While this
isomerization is not as chemically straightforward as the

more classical 20R--->20R+20S sterane or 22R--->22R+22S
hopane epimerization, it can be used as a general indicator of thermal maturity ranking. For this reason, the
14β,17β/(14β,17β + 14α,17α) ratio of the ethylcholestanes
has been calculated (where data are available) for the
grahamites and pyrobitumens of the present study (Table
II).   Results indicate that, while the ββ/ββ+αα ratios
of the grahamites range from 0.45-0.61, and suggest
marginal to full thermal maturity, the single pyrobitumen
for which data are available (AK7446) has a ββ/ββ+αα
ratio of only 0.39, suggesting immaturity. The immature
appearance of the sterane distribution of this Canadian
pyrobitumen can be seen in the top m/z 217 chromatogram
of Figure 7.

Using a second sterane isomer ratio, namely the
20S/20R ethylcholestane epimer ratio, another pyrobitumen
can be shown to be even more immature than AK7446.   The
m/z 217 mass chromatogram for sample AK7445, the Utah
ingramite, is shown at the bottom of Figure 7.

Here, the ββ/ββ+αα ratio is even less than that of
sample AK7446. Even more interesting however, is the
extremely low 20S/20R-5α,14α,17α-ethylcholestane ratio:
approximately 0.2. This value is lower than that for any
crude oil known to the author. This extremely immature
sample, therefore, is almost certainly not the result of
alteration of a conventional crude oil, but rather
appears to have been derived from "extrusion" of organic
matter from a rich, albeit immature source rock[4].
Further, it is clear that the term "pyrobitumen" possesses no genetic connotation whatsoever: while some
pyrobitumens such as impsonites probably result from high

thermal regimes, others such as the Utah ingramite
(AK7445) discussed here originated with very low thermal
input.

4.3 Solid Bitumens -- Source Rock Characteristics

In addition to low thermal maturity, the m/z 217 mass
chromatograms in Figure 7 reveal one other characteristic
that these two pyrobitumens have in common: the almost
total lack of rearranged steranes. The absence of
diasteranes in a source rock or oil is often considered
to imply a lack of smectite family clays (e.g., mont-
morillonite) during diagenesis of the sediments, insomuch
as such clays are thought to catalyze the backbone rearrangement of biochemically-configured steroidal hydrocarbons[27]. Thus the absence of diasteranes has been
noted in oils associated with non-clastic sequences[3], as
such as the oils and rocks of the California mid-
Miocene[6]. The absence of these rearranged compounds in
the two pyrobitumens shown in Figure 7 can, therefore,
tentatively be attributed to a non-clastic source rock.
Such a conclusion is consistent with the current stratigraphic location of the Utah ingramite: this solid
bitumen is found in the calcareous Green River Formation
of the Uinta Basin[15]. Although no definitive solid
bitumen-source rock correlation of the Canadian albertite
has been attempted (however, see ref. 18), it is very
probable that this albertite was also sourced from a
calcareous, or at least low-clastics, source rock.

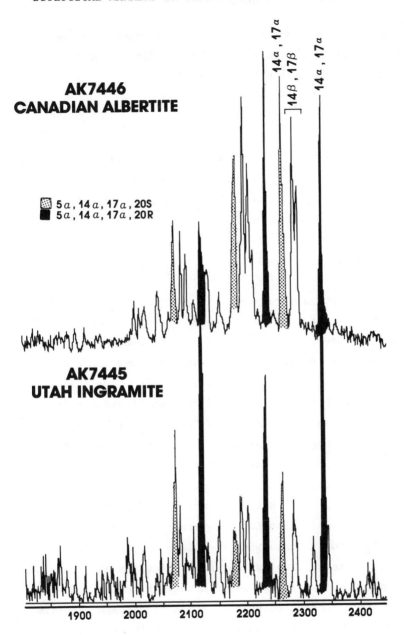

FIGURE 7   M/z 217 mass chromatograms for two pyro-

bitumens; both samples are relatively immature as shown by 14β,17β/14α,17α ratios in the Canadian albertite and 20S/20R ratios in the Utah ingramite.

Thus, knowledge of the extractable biological marker distribution in solid bitumens can be used to assess source rock type, in an analagous manner to the methods utilized for crude oils.

5. SUMMARY

The petroleum explorationist is commonly faced with the question of the genesis of solid bitumens encountered in the well bore or at outcrop. Because grahamites and pyrobitumens are related to both crude oils and crude oil source rocks, biological marker technology applicable to oils and rock extracts has a direct application to the study of these solid bitumens. Conventional biomarker efforts in such areas as organic matter origin, correlation and maturation, can be directly applied to the extractable organic matter fraction of grahamites and pyrobitumens. This study has shown that such efforts can be successful in grouping bitumens into co-genetic families, and in assessing the thermal maturity of pyrobitumens. Biological marker studies can assist in assessing the origin of certain solid bitumens. However, it must be remembered that over 99% of certain pyrobitumens are insoluble in common organic solvents. Thus, when we study the extractable fraction we study only 1%

of the sample, and conclusions about the other 99% must
be inferred from these results. Future efforts at
elucidating the biological marker composition of largely
insoluble grahamites and pyrobitumens should therefore be
directed toward an understanding of this insoluble
fraction, most immediately via pyrolysis-gas chromatography-mass spectrometry. It is only through examining
the total biomarker content that we can expect to fully
understand the chemical history of grahamites and pyrobitumens.

ACKNOWLEDGEMENTS

I wish to thank the following for their contributions to
this study: D. Cameron, D. V. Davis and D. Cardin, for
their assistance with the GCMS data; G. H. Smith, for
contributing several samples, providing supplementary
data, and generously giving me the benefit of his think-
ing concerning solid bitumens and their origin; M. Jacob
and S. A. Bharvani, for laboratory processing of samples,
and M. Jicha for typing. I also extend my appreciation
to J. M. Moldowan for a thorough review of the manu-
script, and to Union Science & Technology Division
management for permission to publish these results.

REFERENCES

1.   K. D. BARTLE, B. FRERE, M. MULLIGAN, S. SARAC AND E.
     EKINCI, Chem. Geol. 34, pp. 151-164 (1981).
2.   J. CONNAN, Advances in Petroleum Geochemistry, v. 1,
     edited by J. Brooks and D. Welte, (Academic Press,
     London, 1984), pp. 299-336.
3.   J. CONNAN, J-L. GRONDIN, J-P. COLIN, G. HUSSLER, and
     P. ALBRECHT, presented at the Eleventh International
     Meeting on Organic Geochemistry, The Hague (1983).
4.   J. A. CURIALE, published with the Proceedings of the
     International Organic Geochemistry meeting in
     Julich, West Germany, in preparation (1985).
5.   J. A. CURIALE, S. R. LARTER, R. E. SWEENEY, and B.
     W. BROMLEY, in Molecular Thermal Maturity Indicators
     in Oil and Gas Source Rocks, Proceedings of the
     Thermal Maturation of Organic Matter Symposium,
     (1985).
6.   J. A. CURIALE, D. CAMERON and D. V. DAVIS, Geochim.
     Cosmochim. Acta 49, pp. 271-288 (1985).
7.   J. A. CURIALE, Petroleum Occurrences and Source-Rock
     Potential of the Ouachita Mountains, Southeastern
     Oklahoma, Oklahoma Geological Survey Bulletin 135,
     p. 65 (1983).

8. J. A. CURIALE, Source Rock Geochemistry and Liquid
    and Solid Petroleum Occurrences of the Ouachita
    Mountains, Oklahoma, unpublished Ph.D. Dissertation,
    University of Oklahoma, 286 p. (1981).
9. J. A. CURIALE, W. E. HARRISON, and G. SMITH, Geo-
    chim. Cosmochim. Acta, 47, pp. 517-523. (1983).
10. J. A. CURIALE and W. E. HARRISON Bull. Amer. Assoc.
    Petr. Geol. 65, pp. 2426-2432. (1981).
11. A. G. DOUGLAS and P. J. GRANTHAM, in Advances in
    Organic Geochemistry, edited by B. Tissot and F.
    Beinner, 1973, pp. 261-276. (John Wiley, 1973).
12. P. J. GRANTHAM, J. POSTHUMA and K. DEGROOT, in
    Advances in Organic Geochemistry edited by A. G.
    Douglas and J. R. Maxwell, pp. 29-38. (Pergamon,
    London, 1981).
13. P. J. GRANTHAM, The Organic Geochemistry of some
    Fossil Resins, Bitumens and Asphalts, unpublished
    Ph.D. Dissertation, University of Newcastle upon
    Tyne, 278 p. (1975).
14. W-Y. HUANG and W. G. MEINSCHEIN, Geochim Cosmochim
    Acta, 43, pp. 739-746 (1979).
15. J. M. HUNT, in Oil and Gas Possibilities of Utah,
    Re-Evaluated, Utah Geological and Mineralogical
    Survey Bulletin 54, pp. 249-273 (1963).
16. H. JACOB, Erdol und Kohle, Erdgas Petrochemie,
    Compendium 76/77, p. 36-49 (1976).
17. S. JI-YANG, A. S. MACKENZIE, R. ALEXANDER, G.
    EGLINTON, A. P. GOWAR, G. A. WOLFF, and J. R.
    MAXWELL, Chem. Geol. 35, pp. 1-31 (1982).
18. G. KHAVARI-KHORASANI, Bull. Can. Petr. Geol. 31, pp.
    123-126 (1983).
19. A. S. MACKENZIE, in Advances in Petroleum Geo-
    chemistry, v. 1, edited by J. Brooks and D. Welte,
    pp.115-214 (Academic Press, London, 1984).
20. A. S. MACKENZIE, L. REN-WEI, J. R. MAXWELL, J. M.
    MOLDOWAN and W. K. SEIFERT, in Advances in Organic
    Geochemistry, Edited by M. Bjoroy et al., pp.
    496-503, (John Wiley & Sons, 1983).
21. W. G. MEINSCHEIN and W-Y. HUANG, in Origin and
    Chemistry of Petroleum, edited by G. Atkinson and J.
    J. Zuckerman, (Pergamon, London, 1981). pp. 33-56.
22. J. M. MOLDOWAN, W. K. SEIFERT, E. ARNOLD, and J.
    CLARDY, Geochim. Cosmochim. Acta, 48, pp. 1651-1662
    (1984).
23. J. RULLKOTTER and D. WENDISCH, Geochim. Cosmochim.
    Acta, 46, pp. 1545-1553 (1982).

24. W. K. SEIFERT and J. M. MOLDOWAN, Geochim. Cos-
     mochim. Acta, 45, pp. 783-794 (1981).
25. W. K. SEIFERT and J. M. MOLDOWAN, Geochim. Cos-
     mochim. Acta, 43, pp. 111-126 (1979).
26. W. K. SEIFERT and J. M. MOLDOWAN, Geochim. Cos-
     mochim. Acta, 42, pp. 77-95 (1978).
27. O. SIESKIND, G. JOLLY, and P. ALBRECHT, Geochim.
     Cosmochim. Acta, 43, pp. 1675-1680 (1979).
28. P. V. F. WILLIAMS and F. GOODARZI, in Organic
     Maturation Studies and Fossil Fuel Exploration,
     edited by J. Brooks, pp. 319-336 (Academic Press,
     London, 1981).

BIOMARKERS AS SOURCE INPUT INDICATORS IN SOURCE ROCKS
OF SEVERAL TERRESTRIAL BASINS OF CHINA

XIANZHANG ZENG, SHUZHEN LIU, SHUNPING MA
Research Institute of Petroleum Exploration & Development, North China
Renqiu County, Hebei Province, China

The composition and distribution characteristics of
steranes and terpanes in sedimentary rocks of several
terrestrial basins of China are analyzed by GC-MS, and
some biomarkers which may be source input indicators
are studied. The relative distributions of C27, C28,
and C29 steranes are indicative of the type of organic
input in Meso-Cenozoic source rocks in these basins.
The resulting characterization of source type mostly
corresponds with kerogen analysis results. Discrep-
ancies may be due to maturation effects on the kerogen
data.

1. INTRODUCTION

It is well known that the oil generation potential of a

source rock is greatly influenced by the type of organic

matrix. Moreover, the composition of the generated hydro-

carbons also depend upon the organic matter. Therefore,

much emphasis is placed on identifying the type of oil-

generative matrix, when an oil/gas-bearing basin is to be

evaluated by geologists.

The type of oil-generative matrix is mainly dependent

on a combination of characteristics inherent in the input

organic matter. The molecular fossils called "biomarkers"
can provide the important information for characterization
of that organic matter. In this regard, Meinschein and
Huang (1981) studied the relative composition of various
sterols in sediments of different sources and in organisms[2].
They suggested that sterol distributions can be used as
source input indicators. Ekweozor et al (1979) found
Oleanane to be a good input indicator of terrestrial higher
plants in crude oil of the Niger Delta, Nigeria[3]. Simoneit
(1976)[4], Philp et al (1981)[5] and Richardson and Mimer
(1982)[6] have made a thorough study of sesquiterpenoids and
diterpenoids in sediments and crude oil and recommended
some biomarkers to be input indicators of terrestrial higher
plants.

      Since 1982, we have performed GC-MS analyses of ster-
anes and terpanes in source rocks of differing matrix type
and in sedimentary environment   We also studied some biomarkers which are characteristic of terrestrial sedimentary
matter in China and can be used as source input indicators.
With such biomarkers, we also made a preliminary attempt to
group the type of organic matter in Meso-Cenozoic source
rocks of several Chinese terrestrial basins[7,8]. This paper
reveals only a part of our research results.

2. EXPERIMENTS

2.1. Samples

Two groups of samples were collected in all.

      The first group was taken from Rao-yang depression in
North China, south of Beijing and Tianjin (Fig. 1). The
famous Renqiu buried hill oil field located there is at the
site of an Oligocene plains-type palaeolake. Liang et al

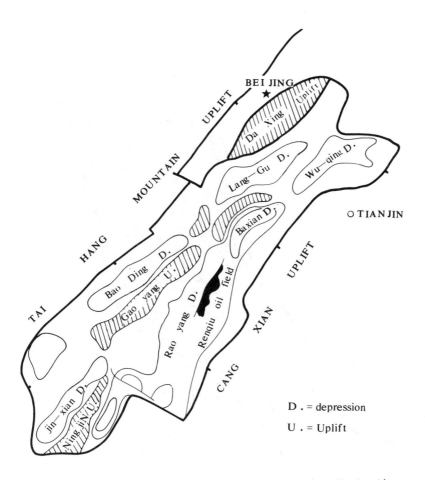

FIGURE. 1   Geographic location of Rao-yang depression, North China

(1982) have made a thorough study of this depression's sedimentary facies and palaeobiofacies, and proposed a division
among the organic facies of its Sa-he-jie formation (Es3)
source rocks (Fig. 2)[9]. We collected 8 samples of source
rocks on various organic facies belts of this depression.
Among them, W32, Ch5 and H3 belonging to the xylophyta
facies are at the margin of the depression, with higher

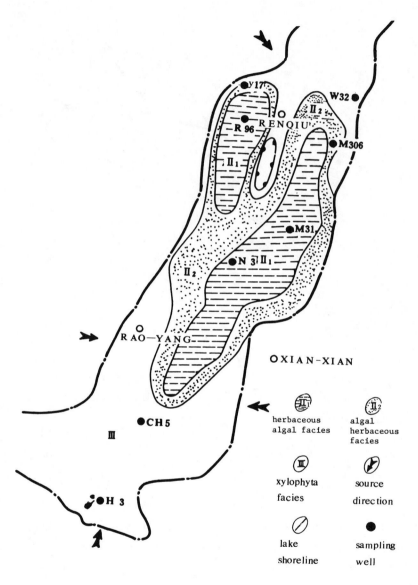

FIGURE 2   Sampling wells and organic facies of Rao—yang depres—
sion during Es₃ stage,   North China

TABLE I Geological and Geochemical Data for the Second Group of Samples

Region	Well Number	Well Depth (m)	Forma-tion	Lithology	Kerogen Analysis Data Atomic H/C	Ratio O/C	type of kerogen	Sedimentary Facies
Song-Liao basin	D111	1664	$K_1Q_n^{2+3}$	black mudstone	1.26	0.107	II_1	deeper lake
Ji-yang depression	W33	1817	Es3	grey mudstone	1.25	0.11	II_1	deeper lake
Shan-gan-ning basin	H45	980	J_1Y	grey-black mudstone	0.71	0.15	III	swamp
Inner-Mongolia Br-lian basin	A3	830	K_1b_2	dark grey mudstone	0.98	0.13	III	fresh water lake
Ji-zhong depression	Y22	3047	O	grey-black limestone	lower aquatic organisms			marine facies
Hebei Tang-shan	/	/	C-P	humic coal	higher plants			swamp
Shan-xi Hun-yuan	/	/	C-P	sapropelic coal	algal is predominant			swamp

plant debris being the predominant organic matters. Source
rocks are of mixed origin, belonging to the algal herbaceous
facies. Source rocks M31, R96 and N3 are taken from the
center of the depression, with lower aquatic organisms being
the predominant organic matter. They belong to herbaceous
algal facies (Fig. 2).

The second group taken from several basins consists of
five source rocks containing diverse types of organic matter
and two coals:

--- Two fine terrestrial source rocks, D111 and W33,
    taken from Song-Liao basin and Ji-yang depression
    of Bo-hai basin respectively;

--- Two poor terrestrial source rocks, H45 and A3,
    taken from Shan-gan-ning basin and Inner-Mongo-lia
    Er-lian basin respectively;

--- One Ordovician marine limestone, taken from Jizhong
    depression of Bo-hai basin;

--- One humic coal of higher plant origin, taken from
    Tang-shan of Hebei province; and

--- One sapropelic coal of algal origin, taken from
    Hun-yuan of Shanxi province.

Table 1 summarizes the relevant geological and geochemical data for the second group.

2.2 Extraction and Separation

Samples were treated as follows:

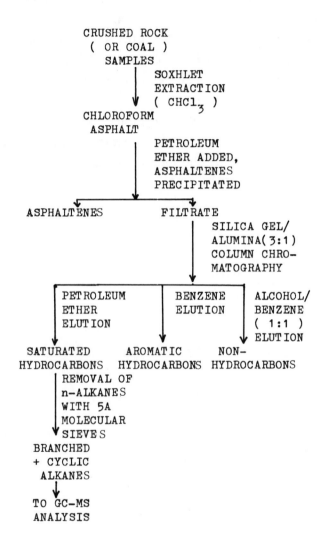

2.3. GC-MS Analysis

GC-MS analyses for the samples were carried out on JGC-20KP/
JMS-D300/JMA-2000 GC-MS-C as follows:

GC --- 30m X 0.25mm OV-101 glass-capillary column,
oven programmed from 80 to 280°C at 2°C/min. He as carrier,
flow rat 3ml/min., injector temperature 310°C, and split
ratio of 100:1.

Interface --- single stage glass jet separator, 280°C.

MS --- EI source, R=500, ionization voltage 70eV, ionization current 300uA, ion source temperature 250°C, scan
mode MID.

3. RESULTS AND DISCUSSION

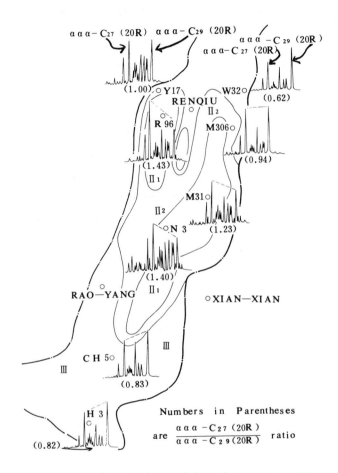

FIGURE 3   Distribution characteristics of steranes in different organic

facies of Rao—Yang depression (Es₃) North China

Analysis results of the first group are shown in Figs. 3 and 4.

As shown in Fig.3, for the source rocks of different organic facies belts at Es3 of Rao-yang depression of North China region, the distribution characteristics of steranes are obviously different.

Near the margin of the depression, the xylophyta facies source rock, with predominant organic matter of terrestrial higher plant debris (W32, Ch5 and H3), is rich in C_{29} steranes (III, refer to Appendix). The relative abundance of C_{27} steranes (I) are low, and $\alpha\alpha\alpha -C_{27}$ (20R)/ $\alpha\alpha\alpha -C_{29}$ (20R) $<$ 1.0 (0.62-0.83).

By contrast, at the center of the depression, the herbaceous algal facies source rock, with predominant sedimentary organic matter of lower aquatic organisms (R96, M31 and N3), is rich in C_{27} steranes. The relative abundances of C_{29} steranes are low, and $\alpha\alpha\alpha -C_{27}$(20R)/ $\alpha\alpha\alpha -C_{29}$ (20R)$>$1.0 (1.23-1.43).

As for the source rocks of the Algal herbaceous facies with mixed origins of organic matter (Y17 and M306), the relative abundances of C_{27} and C_{29} are roughly equal, and $\alpha\alpha\alpha -C_{27}$(20R) / $\alpha\alpha\alpha -C_{29}$(20R) is 0.94-1.0.

Again, as shown in Fig. 4, the distribution characteristics of terpanes are obviously different.

For the xylophyta facies source rocks at the margin of the depression, besides Hopanes(X), there are some triterpanes such as Oleanane(XI) and possibly Spirotriterpanes (XII), Lupane-I(XIII), γ -Lupane(XIV) and Onocerane(XV) to various extents, but the relative abundance of Gammacerane (XVI) is low.

By contrast, most of the herbaceous algal facies

source rocks at the center of the depression contain mainly
Hopanes and Gammacerane.

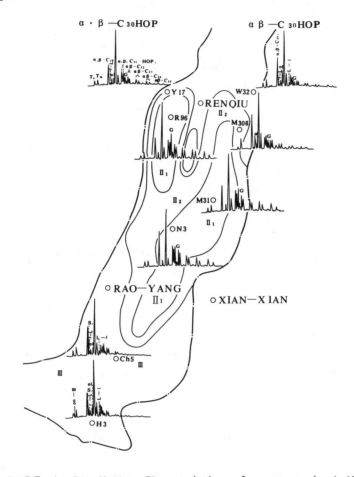

FIGURE 4   Distribution Characteristics of terpanes (m/z 191)
in different Organic facies of Rao—Yang depres-
sion (E s3) , North China.

S : spirotriterpane. r—L : r—Lupane, ol : oleanane. L— 1 : Lupane— I . G : Gammacerane .   on— III : Onocerane — III .
Recognized by comparison of   their mass spectra and
relative retention time with data   of literature[10,11].

As for the algal herbaceous facies source rocks, the distribution characteristics are between the two mentioned above.

Hence, it is seen that the composition and distribution characteristics of sterane and terpane biomarkers can provide important information to determine the source of sedimentary organic matters.

The viewpoint shown above is further confirmed by the analysis results of the second group.

Fig. 5 shows mass fragmentograms for the steranes (m/z 217) of the second group. Both Cretaceous Q_n^{2+3} (D111) of Song-Liao basin and Es3(W33) of Ji-yang depression of Bo-hai basin are type II_1 source rocks which are predominantly sapropelic. The sterane distribution characteristics are similar to the algal coal and Ordovician marine limestone, in which the relative content of C_{27} steranes is as high as 40-46.5% while the relative content of C_{29} is as low as 31% maximum. However, for Jurassic Yan-an formation (H45) of Shan-gan-ning basin and Lower Cretaceous(A3) coal measure source rocks, relative contents of C_{27} are as low as 27-29%, while C_{29} is $>$ 45%. It is similar to the distribution characteristics of humic coal originating from higher plants.

Hence, it is seen that C_{29} steranes could be used as the input indicators of terrestrial higher plants while C_{27} could be used as the input indicators of lower aquatic organisms. Taking into account relative abundance, fine source rocks with good potential could be distinquished from poor ones.

It is interesting that type II_1 source rocks of the Song-Liao basin are relatively rich in 4-Methyl-steranes (IV), while Es3 type II_1 source rocks of Ji-yang depression

of Bo-hai basin show relatively abundant C_{28} (II) steranes.
Such features have not been seen in humic source rocks
(type III) originating

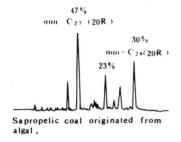

Sapropelic coal originated from
algal .

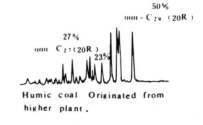

Humic coal Originated from
higher plant .

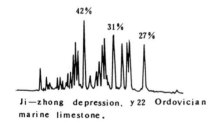

Ji—zhong depression. y 22 Ordovician
marine limestone.

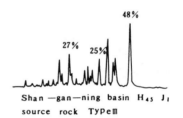

Shan —gan—ning basin H₄₅ J₁
source rock TYPeⅢ

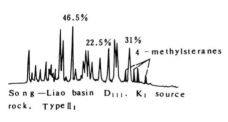

Song —Liao basin D₁₁₁. K₁ source
rock. TypeⅡ₁

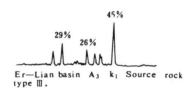

Er—Lian basin A₃ k₁ Source rock
type Ⅲ .

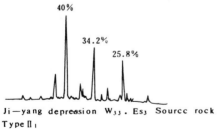

Ji—yang depression W₃₃ . Es₃ Source rock
TypeⅡ₁

FIGURE 5   Distribution characteristics of steranes in several source rocks
           of different type and coal.

mainly from higher plants. So it seems that these compounds also indicate the input of lower aquatic organisms.

Fig. 6 shows the GC-MS analysis of bicyclic-sesquiterpanes and tricyclic terpanes of the second group. Obviously, both humic source rocks from the Jurassic Yan-an formation(H45) of Shan-gan-ning basin, and Lower Cretaceous (A3) of Inner-Mongolia Er-lian basin, and the humic coal contain a relatively high abundance of C_{15} and C_{16} bicyclic-sesquiterpanes (V,VI) and tricyclic diterpanes such as 18-norpimarane (VII) and 18-norabietane (VIII). In the sapropelic source rocks (type II_1, e.g. D111 of Song-Liao basin) which are predominantly of lower aquatic organism, essentially no C_{15} and C_{16} bicyclic-sesquiterpanes could be detected. The dominant tricyclic-sesquiterpanes could be detected. The dominant tricyclic terpanes have a base peak of m/z 191(IX). The relative content of tricyclic diterpanes such as 18-norpimarane and 18-norabietane is quite low.

As for the Lower Palaeozoic Ordovician marine limestone (Y22) deposited before the appearance of higher plants and sapropelic coals which originated from algae, neither C_{15} nor C_{16} bicyclicsesquiterpanes nor tricyclic-diterpanes such as 18-norpimarane and 18-norabietane were found. Only one series of tricyclic terpanes with a base peak of m/z 191 is present.

These results indicate that both C_{15} and C_{16} bicyclic-sesquiterpanes and tricyclic-diterpanes such as 18-nor-pimarane and 18-norabietane can be used as indicators of higher plant input.

Fig. 7 shows the GC-MS analysis of pentacyclic triterpanes of the second group. The humic source rocks (H45 of Shan-gan-ning basin and A3 of Er-lain basin) are similar

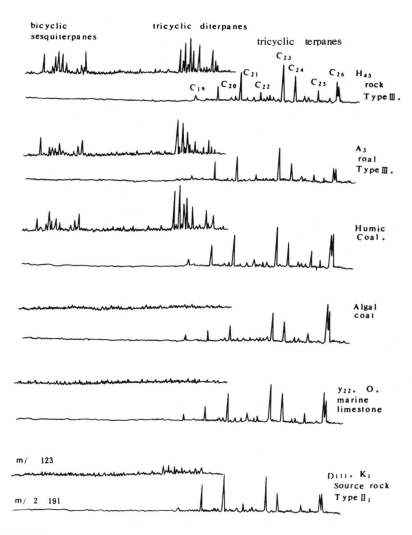

FIGURE 6   Mass fragmentograms of bicyclic , tricyclic terpanes (m/ z
12 3, m/ z 191)   in several source rocks of different ty—
pes and coal.

Recognized by comparison of their mass spectra and re—
lative retention time with data of literature. 5.2

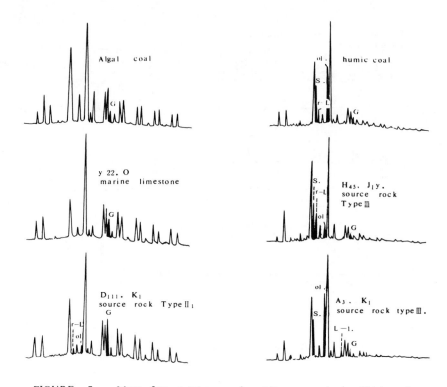

FIGURE 7   Mass fragmentograms of   triterpanes   (m/z 191)   in
             several   source   rock   of   different   type   and   coal
             S: spirotriterpane   r—L: r—lupane   ol: oleanane
             L—1: Lupane—1   G: Gammacerane.
             Recognized by Comparison of their mass spectra and
             relative   retention   time with data of literature[10, 11]

to humic coal in their pentacyclic triterpane patterns.
Besides hopanes, other terpanes such as oleananes, spiro-
terpanes and γ-Lupanes(XIV) are relatively abundant.

     In the type II_1 sapropelic source rock (D111 of Song-
Liao basin), such triterpanes (e.g. oleanane and spiroter-
pane), are not detectable, but hopanes and gammacerane are

predominant. In the Ordovician marine limestone and the algal coal, pentacyclic **triterpanes** other than hopanes were not found.

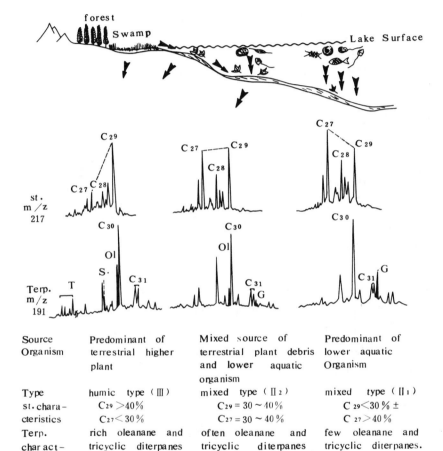

Source Organism	Predominant of terrestrial higher plant	Mixed source of terrestrial plant debris and lower aquatic organism	Predominant of lower aquatic Organism
Type st. characteristics	humic type (Ⅲ) $C_{29} > 40\%$ $C_{27} < 30\%$	mixed type (Ⅱ₂) $C_{29} = 30 \sim 40\%$ $C_{27} = 30 \sim 40\%$	mixed type (Ⅱ₁) $C_{29} < 30\% \pm$ $C_{27} > 40\%$
Terp. characteristics	rich oleanane and tricyclic diterpanes few gammacerane.	often oleanane and tricyclic diterpanes existed with moderate gammacerane.	few oleanane and tricyclic diterpanes. rich gammacerane

FIGURE 8   Source Organism of   different facies in   terrestrial   lake basin and   their characteristics   steranes   and   terpanes.

St: Sterane   T: tricyclic diterpane   Ol: oleanane

Terp: terpane   S: spirotriterpane   G: Gammacerane.

This fully indicates that the precursors of oleananes, spirotriterpanes and γ-lupanes originate from terrestrial higher plants, and they could be used as terrestrial input indicators, while gammacerane could be used as an input indicator of lower aquatic organisms.

In summary, the composition and distribution of steranes and terpanes in various types of organic matter found in source rocks are illustrated in Fig. 8.

4. APPLIED EXAMPLES

Based on the above analytical results, we apply the relative compositions of three steranes (i.e. $\alpha\alpha\alpha$ -C_{27}-20R, $\alpha\alpha\alpha$ -C_{28}-20R and $\alpha\alpha\alpha$ -C_{29}-20R) to make a division of sources and types of organic matter for 61 source rocks in several major terrestrial oil/gas-bearing basins of China (see Fig. 9).

As shown in Fig. 9, the source matter of the Cretaceous Qing-shan-kou formation Q_n^{2+3} of Song-Liao basin, the Oligocene Sha-he-jie formation of Ji-yang depression of Bohai basin, and the Sha-he-jie formation of Rao-yang depression of North China are all predominantly type II_1, indicating that their organic matter originated mainly from lower aquatic organisms.

The source rocks of the Sha-he-jie formation of Zhong-yuan Dongpu depression, Ji-zhong Shen-xian depression and Ji-zhong Lang-Gu depression are predominantly type II_2, indicating that their organic matter is from mixed sources.

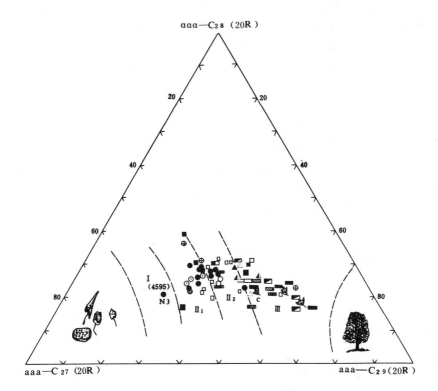

FIGURE 9   Ternary diagram showing the relative constituent of
steranes in different type of source rock

●	Rao—yang depression Es	◍	shenxian depresson. E.
⊕	Ji—yang depression, Es	⊙	Song—Liao basin K₁
○	Lang—Gu depression. Es	△	Qiuxian depression. K₁
▲	Shijiazhuang depression K₁	▨	Er—Lian basin K₁
⧊	Beijing depression, K₁.	◨	zhungeer basin C —T
▰	Shan—gan—ning basin, J₁	□	Dong—pu depression. Es.
■	Nan—xiang basin, Es.	⚘	Baoding depression. Ed—Es₁
(4595)	buried depth, m.		

The source rocks of the Jurassic Yan-an formation of Shan-gan-ning basin, the Permian, Triassic and Carboniferous of Xinjiang Zhungeer basin, the Lower Cretaceous of Inner-Mongolia Er-lian basin, and Sha-he-jie formation of Ji-zhong Bao-ding depression are essentially type III, indicating that the predominant organic matter is terrestrial higher plant.

This resulting division is mostly in correspondence with the kerogen elemental data (Fig. 10). In a total of 61 samples, classification by kerogen elemental analysis data for 51 source rocks corresponds with the divisions made by steranes, accounting for 83.6%. A few samples did not correlate, mainly the source rocks which are buried

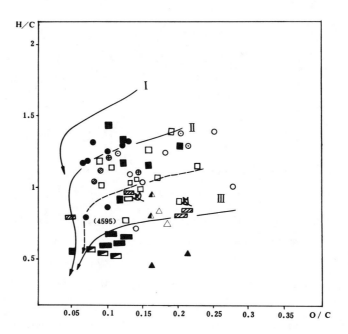

FIGURE 10   Elemental composition of Kerogens for   Source rock
in Fig. 9

deeper and highly matured. For example, N3 of Rao-yang depression is the source rock buried at 4594m. It would belong to type I based on the sterane ternary diagram, but
would belong to type II_2 based on the kerogen elemental
analysis data. We believe this difference results from
influence of maturity.

It is well known that for source rocks of low maturity
it is feasible to distinguish the types of organic matter
with kerogen elemental analysis data. However, as maturity
increases, kerogen deteriorates, deoxidizes and dehydro-
genates, until completely degraded. Therefore, for those
source rocks with higher maturity, classifications accord-

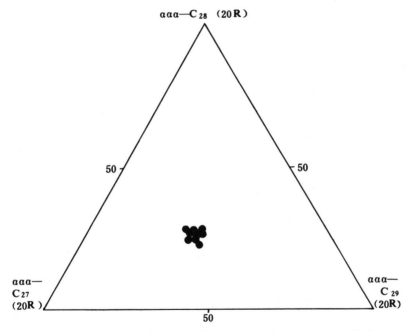

FIGURE 11   Ternary diagram of steranes in source rock during
thermal simulation (by Zhou Guaingjia et al. 1984)

ing to kerogen elemental analysis are thought to be in error. On the other hand, the relative composition in the ternary diagram of steranes is less influenced by maturity. The source rocks with higher maturity still reflect the compositional characteristics of the "original" organic matter. This has been confirmed by laboratory thermal simulation experiment[13]. Fig. 11 shows the result of a thermal simulation experiment on one immature rock sample (Ro = 0.35%) under varying temperatures. It indicates that the relative composition of steranes will not change much even if the maturity increases up to Ro = 1.43%. Therefore, an important advantage of independence from maturity effects is evident in using the sterane ternary diagram to identify the source of the organic matter and to classify samples into types.

5. CONCLUSION

Through analysis of two groups of samples taken from different organic facies belts having different types of organic matter, the following two conclusions are drawn:

1. The composition and distribution characteristics of biomarker steranes and terpanes can provide important information about source of organic matter in source rocks. Those that can be used as input indicators of terrestrial higher plants are 24-ethyl-C_{29} sterane, C_{15} and C_{16} bicyclic-sesquiterpanes, tricyclic diterpanes such as 18-nor-pimarane and 18-norabietane, and non-Hopane triterpanes such as Oleanane, Spirotriterpane, and γ-lupane. C_{27} and C_{28} steranes, 4-methyl-steranes, Gammacerane and Hopanes can be used as input indicators of lower aquatic organisms.

2. The relative composition ternary diagram of three identical configurations of steranes, $\alpha\alpha\alpha$-20R-C_{27},-C_{28},

and $-C_{29}$, is applicable to identification of the sources of sedimentary organic matter and can be used to classify source rocks by source type. This method has an important advantage of independence from the influence of maturity.

REFERENCES

1. M.A. Rogers, Application of Organic Facies Concept to Hydrocarbon Source Rock Evaluation. 10th. World Petroleum Congress PD1, 1979
2. W.G. Meinschein and W.Y. Huang, In: Origin and Chemistry of Petroleum (G. Atkinson and J.J. Zuckerman eds) pp. 33–55, Pergamon Press, 1981
3. C.M. Ekweozor, J.I. Okogun, D.E.U. Ekong and J.R. Maxwell, Chemical Geology, 27, 11–28 (1979)
4. B.R.T. Simoneit, Geochim. Cosmochim. Acta, 41, 463–476, (1977)
5. R.P. Philp, T.D. Gilbert and J. Friedrich, Geochim. Cosmochim. Acta. 45, 1173–1180 (1981)
6. J.S. Richardson and D.E. Mimer, Anal. Chem. 54, 765–768 (1982)
7. X. Zeng, Z. Wang, X. Zhang, X. Chen, and S. Liu, Distribution Characteristics and Geological Significance of Biomarkers in Crude Oil and Source Rocks of Jizhong Depression, Abstract of 1st National Meeting on Organic Geochemistry, Society of Sedimentology, Chinese Society of Mineralogy, Petrology and Geochemistry (Guiyang, China, 1982) pp. 129–130.
8. X. Zeng, S. Liu, S. Ma, J. Wang, Z. Wang, Several Molecular Fossils Being as Input Indicators of Sources, Papers Abstracts of 2nd National Meeting on Organic Geochemistry and Terrestrial Oil Genesis, Chinese Petroleum Society and Chinese Society of Mineralogy, Petrology and Geochemistry, 1984
9. D. Liang, X. Wang, L. Mou, B. Liu, X. Zeng, G. Duan, and Q. Luo, Genesis of Oil/gas in Rao-yang Depression, Int. Collection of Research Papers of Exploration and Development in North China Oilfield, Research Institute of Petroleum Exploration & Development of North China (1982), pp. 1–40.
10. B.J. Kimble, J.R. Maxwell, R.P. Philp and G. Eglinton, Chemical Geology, 14, pp. 173–198 (1974)
11. I.R. Hills and E.V. Whitehead, Nature, Vol. 209, No. 5027, pp. 977–979 (1966)

12. R.C. Barrick and J.I. Hedges, Geochim. Cosmochim. Acta Vol. 45 No. e, pp. 381-392 (1981)

13. G. Zhou, S. Li, Z. Chen, and Y. Song, Discussion on Index of Common Biomarkers of Terpanes and Steranes, Collection of Research Papers of Exploration and Development in Shengli Oilfield (Petroleum Geochemistry Issue), Research Institute of Geological Science of Shengli Oilfield (1984)

Appendix

A NEW TECHNIQUE FOR OIL/SOURCE-ROCK AND OIL/OIL CORRELATION.

F. BEHAR, R. PELET
Institut Français du Pétrole
Géologie et Géochimie
Rueil-Malmaison, France

Asphaltenes derived from both oil and source-rocks, and kerogens, have been studied by pyrolysis-GC and GC-MS. It has been shown that the comparison between pyrolyzates of asphaltenes from oils and asphaltenes or kerogen from source-rocks, are valuable tools for correlation purposes. For oil/oil correlations, it can supplement and even replace in the case of degraded oils, the information given by hydrocarbon distribution in the maltene fraction. For oil/source-rock correlations, the match between pyrograms of asphaltenes from both oil and source-rock offers more direct possibilities than usual techniques. This is especially true for the overall comparison of hydrocarbons (GC/ traces). In the detailed analysis of polycyclic saturates (GC/MS), a second order effect appears which can obscure the conclusions. This effect is due to the great sensitivity of polycyclics to thermal evolution, which mandates selection of source-rock samples at a sufficient level of evolution. Another drawback is the lack of specificity of pentacyclic distributions, which makes them unsuitable for correlation purposes.

1. INTRODUCTION

Correlating oils to oils and oils to source-rocks is the first step in the geochemical study of a sedimentary basin. Correlating oils to oils serves to condense numerous

samples into a few groups, each having a distinct origin
(= source-rock). This origin can be assessed by correlating
oils to source-rocks,and more precisely,to some organic
extract fractions of the source-rocks . These correlations
are accomplished by matching some compositional pattern(s)
which can belong to two types: bulk (e.g. ^{13}C) or
semi-detailed (e.g. GC of n-alkanes) parameters,
representative of an important fraction of the materials
under investigation, and very detailed (e.g. GC/MS of
hopanes/moretanes) analysis for a restricted range of low
content compounds possessing a highly characteristic
structure ("biomarkers").

Both types look very suitable for oil/oil
correlations. However, when reservoired oils have suffered
deep modifications (e.g. microbial degradation) (CONNAN
1984), the first type cannot be used and the second must be
used with great care.

For oil/source-rock correlations, the first type
suffers from internal inconsistence : important
compositional variations take place during primary
migration, i.e. when mobile compounds are expelled from the
source-rock to form reservoired oils (VANDENBROUCKE 1972,
DURAND and OUDIN 1981, LEYTHAEUSER et al 1981, 1984). Thus,
whatever the parameter used, perfect matches cannot be
expected and, moreover, must even be avoided. The rules
governing those compositional variations are not
quantitatively known to date. But even if they were known,
they would obviously depend on so many factors (LEYTHAEUSER
et al 1984) that the precise determination of the
compositional variations would remain practically extremely
difficult. These limitations do not apply to the second type
of analysis : compounds of the same overall structure in a

limited range of carbon numbers will have practically the
same fate during primary migration. But the specificity and
the representativity of those compounds is questionable.
Because of the lack of fundamental studies, specificity must
be checked empirically, by the collection of available data.
For example, it is a common observation that the so-called
"normal" distribution of hopanes remains very similar for a
great number of oils and rock extracts without any genetic
relationship. This phenomenon is apparent for the samples
used in this work. However, the most important problem is
representativity. It arises from the low content of the
compounds under study. The process of oil accumulation in
pools is homogenizing by its very nature ; thus the
distribution of biomarkers in a reservoired oil will not
vary significantly at least within the pool limits. But the
formation of this pool requires the drainage of a
source-rock volume many times larger. In this very large
volume, small spatial variations in organic supply and/or
environmental conditions may occur during sedimentary
deposition. Their influence on the bulk composition of
organic matter will be small and even negligible, but they
may perfectly replace locally a biomarker distribution by
another completely different one. It is only on an average
basis that a source-rock biomarker distribution can be
 assessed and then compared to the oil distriution Not
only is this never done, but it is generally impossible, due
to the lack of samples and of detailed sedimentological
knowledge.

        These considerations emphasize the importance of
the first type: bulk parameter correlations. The technique
proposed in this work enables such correlations. free from

migration effects. It enables also bulk correlations on reservoired oils, whatever the secondary alteration they have suffered.

2. PRINCIPLES OF THE METHOD

2.1.   It is proposed to use for correlation purposes the pyrolyzates from asphaltenes of oils and source-rock extracts; for source-rocks, the asphaltene pyrolyzate may be replaced by the kerogen pyrolyzate. Due to mineral matrix effects (ESPITALIE et al 1984) it is not recommended to use in this latter case whole extracted rock pyrolyzates.

2.2.   For oil/oil correlation, it has been already shown that oil asphaltene pyrolyzates have a global and semi-detailed composition close to the composition of the remainder of the oil (RUBINSTEIN et al 1979, AREF'YEV et al 1982, BEHAR et al 1984). Thus, the pyrolysis of asphaltenes from heavily biodegraded oils entirely devoid of n-alkanes generates new n-alkanes close in distribution to those of the corresponding non-biodegraded samples, if available. The reasons for this closeness are discussed in the following section. Identity must not be expected primarily because of pyrolysis artifacts (e.g. unsaturates formation). But if we compare asphaltene pyrolyzates to each other, pyrolysis effects are identical, and even the identical distributions may be expected for correlating oils. Examples are given in the experimental section.

2.3.   The same reasoning could be applied to oil/source-rock correlations if oil asphaltenes and the corresponding source-rock asphaltenes were identical, i.e. asphaltenes would not undergo migration effects. The reverse is obviously true : source-rock extracts contain at least three times more asphaltenes than the oils (TISSOT and WELTE 1978), which calls for a relative retention in the rock during primary migration. This retention has been directly observed in at least one case (VANDENBROUCKE 1972). Now the causes for this retention may be twofold :

- a size effect, the bulkier molecules being retained;
- and/or a polarity effect, the more functionalized molecules being retained

Size fractionations of asphaltenes by GPC have been performed ; although their interpretation in terms of absolute molecular size is questionable, they show that the size distributions extend over a range of at least 2 orders of magnitude (from 1000 to 100000 Dalton). A careful and rather extensive study of 14 size fractions (GOURLAOUEN 1984) shows a continuous and modest variation of their physicochemical properties. This proves that, if there are functional differences between asphaltene molecules, they are largely independent of their size Finally, the failure of fractionation schemes based upon functionality (MONIN and PELET 1984) indicates that functional differences between asphaltene molecules are small, if any. From all these results the following statement can be inferred : asphaltene molecules are composed of various building units; the total number of these units varies greatly, but the relative number of each different unit is statistically the same, and their mutual arrangement is statistically equivalent. Thus, asphaltenes appear as macromolecular, nonpolymeric but statistically homogeneous compounds. This model will result in an expected indentity of asphaltene pyrolyzates from an oil and its corresponding source-rock, provided that, after

oil pooling, the reservoired oil and the source-rock were
not submitted to divergent thermal conditions.

The reality of asphaltene thermal evolution
under natural conditions has been documented by CASTEX
(1985). The greatest evolutionary range found was for the
Douala basin; from the shallowest to the deepest sample of
source-rock, H/C decreased from 1,53 to 1,10 and O/C from
0,20 to 0,10. This is to be compared to the range of
variation of H/C (from 1,18 to 1,13) in the 14 fractions of
the same asphaltenes (GOURLAOUEN 1984 - O/C was not
reported, but was said to fluctuate within the limits of
experimental error). Thus, in order to be correlated oils
and source-rocks must be taken as far as possible, in
equivalent basinal structural situations. Even in that case,
subtle effects remain which are discussed elsewhere (BEHAR
and PELET 1985).

Examples of oil/source-rock correlations are
given in the experimental section.

2.4.   It has already been shown (BEHAR et al 1984)
that the pyrolyzates of asphaltenes and kerogen from the
same source-rock match very closely. They can therefore
replace one another. It is recalled here that this property
is a consequence of the structural similarity of asphaltenes
and kerogen; kerogen molecules being bulkier (thence
insoluble) and more aromatic (the same side chains, bridges,
and functions being fixed on larger aromatic nuclei), and
presenting the same statistical homogeneity. These
similarities are interpreted as a consequence of the common
origin of the whole organic content of a rock, the concept
of PFEIFFER and SAAL (1940) of oils as a continum from the

lightest saturates to the heaviest asphaltenes being extended to the totality of sedimentary organic matter (physically described by an analogue of the coal micellar model of KREULEN 1948). The origin of asphaltenes can be threefold :

        - They are a definite class of size of the organic molecules originally sedimented (smaller than kerogen, larger than resins);

        - They are the heavy ends of resin dismutation (the light end being hydrocarbons) during geological evolution;

        - They are an intermediate in kerogen dismutation (to hydrocarbons on the light end and carbonaceous residues on the heavy end) during geological evolution.

The first origin is quantitatively important only during early diagenesis. We suggest that the second origin may have some relative importance during diagenesis stricto sensu. The overall most important origin is certainly the third one.

        It is clear that the statistical intramolecular homogeneity of asphaltenes does not exist for hydrocarbons which compose building units or fragments of asphaltenes. But when the distribution of the naturally produced fragments (the hydrocarbons) is compared to the distribution of pyrolytically produced fragments, this statistical homogeneity is reestablished, as far as pyrolysis simulates natural evolution. Thence, the result is a closeness of oil alkane distributions, on the one hand, and of oil asphaltene, source-rock asphaltene, and kerogen pyrolyzates alkanes + alkenes, on the other hand (BEHAR et al 1984).

3. **EXPERIMENTAL**

Prior to pyrolysis, kerogen and asphaltenes from source-rock and crude oil are prepared according to the analytical procedure described previously (BEHAR et al. 1984).

Pyrolysis is performed using a minifurnace which was designed in our laboratory (BEHAR et PELET 1985). Kerogens and asphaltenes are heated from 320°C to 550°C at 60°C/mn then left at 550°C for 5 mn. The pyrolyzate is recovered and fractionated by liquid chromatography into saturates + unsaturates, aromatics and polar compounds. The fraction containing saturates + unsaturates is hydrogenated under P_{H_2} 7 MPA for one hour at 120°C with rhodium $Al_2 O_3$ (5%) as catalyst. It is then fractionated with 5 Å molecular sieves into n alkanes and iso-cyclo alkanes. Iso-cyclo alkanes are analyzed by GC/MS.

4 **SAMPLES**

The crude oils and source-rocks come from a Venezuelan basin. Previous studies (unpublished data) have shown that organic matter in source-rocks could be divided into 2 types labeled as A and B. For crude oils, there is a great variation of compositions; one may have either two distinct types also labeled A and B, or compositional intermediates. The latter are considered to be reservoired mixtures in various proportions of the two types A and B. Those two types are considered as generated from two distinct source-rocks. In addition there are also biodegraded oils, difficult or impossible to classify.

In this work, we have used three source-rocks A,

one source rock B and three oils: one A, one B and one biodegraded. Pyrolysis and analysis were carried out on the three rocks A in order to follow kerogen evolution with depth. Consequently, correlations were made as follows :

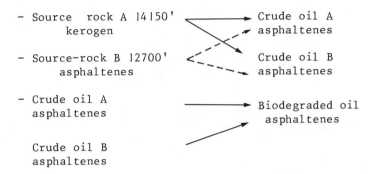

- Source rock A 14150'
  kerogen

- Source-rock B 12700'
  asphaltenes

- Crude oil A
  asphaltenes

  Crude oil B
  asphaltenes

  → Crude oil A
    asphaltenes

  → Crude oil B
    asphaltenes

  → Biodegraded oil
    asphaltenes

Kerogen was taken instead of asphaltenes for rock A because of their low content in the rock extract. It was proved in section 2.4 that this replacement was justified.

5 **RESULTS**

5.1. Kerogen evolution with depth

Pyrolysis was carried out on 3 kerogens A from depths of 5000', 11500', 14500'.

5.1.a. GC Traces

The pyrograms obtained for each kerogen are given in Fig. 1. The alkane + alkene distribution changes with depth: the $C_{25}-C_{30}$ fraction, pristene, and polycyclic molecules in the $C_{25} - C_{35}$ range decrease whereas the $C_{15} - C_{25}$ fraction increases. Of course, these properties are retained after hydrogenation (Fig. 1), although some loss of light ends due to evaporation is apparent. These points are consistent with saturates evolution already observed in rock extracts (TISSOT and WELTE 1984).

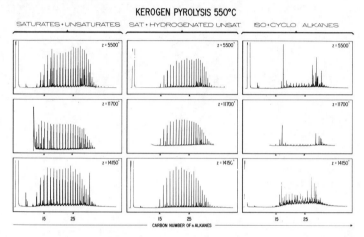

Fig. 1 : Pyrograms of saturates + unsaturates, saturates + hydrogenated unsaturates, iso + cyclo alkanes for three kerogens, from type A source-rocks at 5500', 11700', 14150', depth

5.1b. Structural analysis by CG/MS

Polycyclic molecules were analyzed by GC/MS in order to determine the presence of tri, tetra and pentacyclic biomarkers.

- Tricyclic and pentacyclic structures (m/z = 191). The fragmentation m/z = 191 permits the recognition of two series of tricyclic terpanes (AQUINO NETO et al 1983, EKWEOZOR et al 1983, BEHAR and ALBRECHT 1984) and of pentacyclic triterpanes (VAN DORSSELAER 1975).

For the three kerogens there are no tricyclic structures (Fig. 1).

In the patterns of pentacyclic structures $C_{27}-C_{30}$ molecules predominate as previously reported by SEIFERT (1978),RULLKOTTER et al (1984). BEHAR and PELET 1985). Now, for each carbon number, all of these compounds appear as a mixture of conformational isomers. Fig. 2 and table 1 show the variation of the series $17\beta- 21\beta$ (H), $17\beta -21\alpha$ (H), $17\alpha -21\beta$ (H) with depth. There is a continuous decrease of $17\beta-21\beta$ and $17\beta-21\alpha$ molecules as $17-21$ compounds increase.For the diasteromers 22 S and 22 R in molecules above C_{30}, the ratio S/R also increase with thermal evolution.

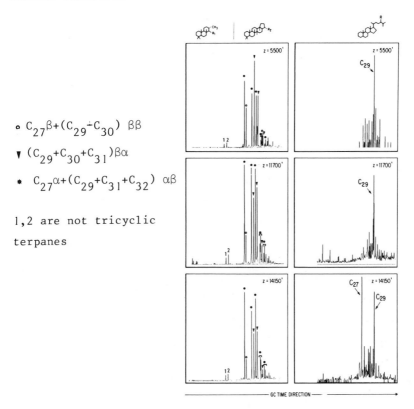

○ $C_{27}\beta+(C_{29}+C_{30})$ $\beta\beta$

▼ $(C_{29}+C_{30}+C_{31})\beta\alpha$

∗ $C_{27}\alpha+(C_{29}+C_{31}+C_{32})$ $\alpha\beta$

1,2 are not tricyclic terpanes

Fig. 2 : Fragmentograms m/z = 191 and m/z = 217 of iso-cyclo alkanes, from three kerogens at 5500', 11700', 14150' depth for type A organic matter.

Sample	$C_{27\beta}+((C_{29}+C_{30})_{\beta\beta}/$ $C_{27\alpha}+$ $(C_{29}+C_{32})_{\alpha\beta}$	$(C_{29}+C_{30})_{\beta\alpha}/$ $(C_{29}+C_{30})_{\alpha\beta}$	$(C_{31}+C_{32})R/$ $(C_{31}+C_{32})S$
Kerogen 5500'	0,43	1,31	0,60
Kerogen 11700'	0,10	0,74	0,67
Kerogen 14150'	0,098	0,62	1,28

Table 1 : Evolution with depth of pentacyclic terpanes
derived from kerogen pyrolysis

Tetracyclic structures (m/z = 217)

Steranes can be studied by monitoring the ion of
$m/z = 217$. The corresponding fragmentograms given in Fig. 2
show also a variation of the molecular distribution with
depth. Up to 11700' we can identify only C_{29} steranes. At
greater depth a mixture of 2 groups of steranes appears,
with 27 and 29 carbon atoms.

5.1.c Discussion

Pyrograms of saturates and unsaturates released
from kerogen pyrolysis show a continuous variation of
molecular distribution which parallels the well-known
evolution of saturates found in rock extracts (TISSOT et
WELTE 1984, HUNT 1979, MACKENZIE et MACKENZIE 1983, SEIFERT
et MOLDOWAN 1980). These results suggest that molecules
bound to kerogen can undergo thermal maturation effects
during burial, without release from the kerogen. Thus,
kerogen pyrolysis effluents can be also used as a maturity
parameter of source-rocks.

A direct implication of the results is in the
choice of the samples for correlation studies. In fact, it

will be necessary to correlate source rock of the same maturity level as the asphaltenes of the corresponding oil. For this reason, we were obliged to take the two source-rocks as deep as possible, 14150' for kerogen A and 12700' for kerogen B.

5.2. Oil source-rock correlation

5.2.a. GC traces

In Fig. 3 are shown pyrograms of asphaltenes derived from both source-rock and crude oil B, in Fig. 4 are given those from kerogen A and crude oil A asphaltenes. For each fraction, saturates + unsaturates, saturates + hydrogenated unsaturates, branched and cyclo alkanes, the oil source-rock correlations are excellent except for the isoprenoïds.

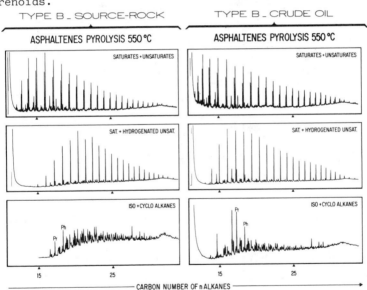

Fig. 3 : Correlation between pyrograms of saturates + unsaturates, saturates + hydrogenated unsaturates, iso-cyclo alkanes for asphaltenes derived from both source rock B and crude oil B.

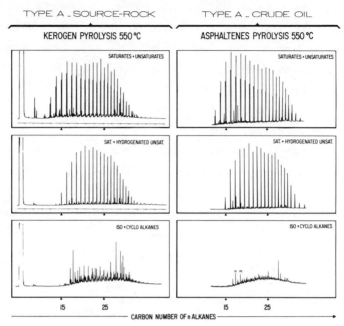

Fig. 4 : Correlation between pyrograms of saturates + unsaturates, saturates + hydrogenated unsaturates, iso-
cyclo alkanes for kerogen A (14150') and asphaltenes
derived from crude oil A.

5.2.b. GC-MS analysis

- Tricyclic and pentacyclic structures (m/z = 191)

The distributions of tricyclic molecules are very
different for the two types A and B kerogens as shown in
Fig. 5 and 6. In fact, Type B contains numerous molecules
which are not present for type A. Furthermore there is a
close similarity between source-rock B and crude oil B, but
for crude oil A tricyclics, present in a low content, show
type B distribution. It was said before (cf. 4) that a
number of crude oils in this basin, are a mixture of the two
types A and B; these results suggest that there is no "pure"
crude oil A i.e. not without a small fraction derived from
type B source-rock.

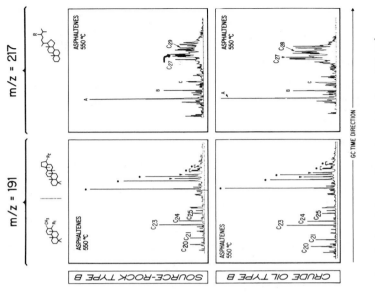

Fig. 6 : Correlation between framentograms m/z = 191 and m/z = 217 of iso-cyclo alkanes for kerogen A (14150') and asphaltenes derived from crude oil A.

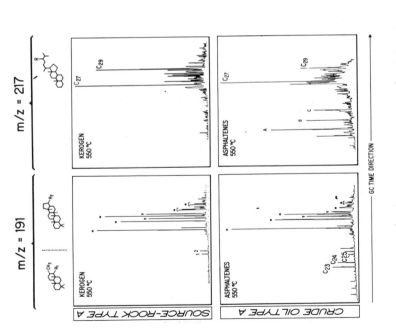

Fig. 5 : Correlation between fragmentograms m/z = 191 and m/z = 217 of iso-cyclo alkanes for asphaltenes derived from both source rock and crude oil B.

Pentacyclics show essentially the two series of 17α -21β and 17β $-$ 21α hopanes with a predominance of molecules ranging from C_{27} to C_{30}. The distributions observed for source rock B and for both crude oils A and B are similar. On the other hand, in kerogen A there are some differences which can be interpretated by a lower level of maturity for kerogen A relative to crude oil asphaltenes A (cf. 5.2.). The distributional similarity of types A and B makes the use of pentacyclic alkanes unsuitable for correlation purposes (cf. Section 1).

- Tetracyclic structures (m/z = 217)

Fragmentograms show distinctive patterns for the two types of organic matter. For type B, steranes are characterized by molecules ranging from C_{27} to C_{29}. Furthermore, the correlation between asphaltenes derived from both source rock and crude oil is obvious. For type A some differences appear between the source-rock and the oil. In fact, C_{27} anc C_{29} molecules are the major compounds in the kerogen whereas C_{27} are much more greater than C_{29} in the asphaltenes pyrolysate of the oil (Fig.6). This can be related to a difference in the maturity level between the two samples, exactly as it has been seen for pentacyclic structures.

The pregnane series (peaks A, B, C; Fig. 5 and 6) present with a medium content in both source rock and oil B is also found in small quantity in oil A. This suggests once again that, as it was deduced for tricyclic structures crude oil A may contain some fraction derived from source rock B.

5.3. OIL-OIL CORRELATIONS

5.3.a. GC analysis

Three different Venezuelan oils are studied, labeled as A, B, biodegraded. Their saturate GC traces are shown in Fig. 7 and are compared with asphaltenes saturates + unsaturates and saturates + hydrogenated unsaturates pyrograms. For each oil A and B, close similarities can be noticed for the three traces. Thus this relationship between hydrocarbons in asphaltene pyrolyzate and hydrocarbons in the parent oil suggests that the pyrolyzate traces for the "biodegraded" sample should mimic the now lacking distribution of saturates. Then the comparison with oils A and B shows that the biodegraded sample is related to type B.

For branched and cyclo-alkanes, the similarity of pyrograms between type B and the biodegraded oil is striking as is the dissimilarity with the pyrogram of type A. This confirms the indentification of the biodegraded oil.

5.3.d. GC-MS analysis

In Fig. 8 are given the fragmentograms $m/z = 191$ and $m/z = 217$ for branched and cyclo alkanes derived from asphaltenes pyrolysis. The distribution of tricyclic molecules in the biodegraded oil shows an excellent correlation with the distribution from type B oil.

On the other hand, the pentacyclic traces are very similar for the three oils, so again they cannot be used for correlation studies.

The similarities in sterane distributions between type B and biodegraded oils confirm the common origin of both samples.

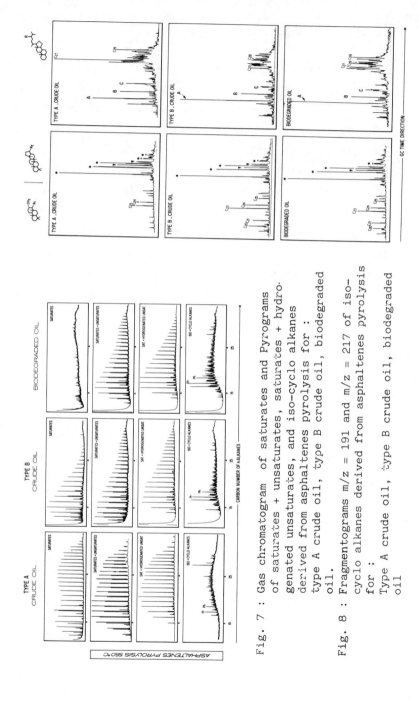

Fig. 7 : Gas chromatogram of saturates and Pyrograms of saturates + unsaturates, saturates + hydrogenated unsaturates, and iso-cyclo alkanes derived from asphaltenes pyrolysis for : type A crude oil, type B crude oil, biodegraded oil.

Fig. 8 : Fragmentograms m/z = 191 and m/z = 217 of iso-cyclo alkanes derived from asphaltenes pyrolysis for : Type A crude oil, type B crude oil, biodegraded oil

CONCLUSION

It has been shown that the comparison between pyrolyzates of asphaltenes from oils, and asphaltenes or kerogen from source-rocks, are a valuable tool for oil/oil and oil/source-rock correlation purposes. This is especially true for the overall comparison of hydrocarbons (GC traces). When one comes to the detailed analysis of polycyclic saturates (GC/MS), a second order effect appear, which can obscure the conclusions. This effect is due to the great sensitivity of polycyclics to thermal evolution, which demandes the selection of source-rock samples at a sufficient level of evolution. Another drawback is the lack of specificity of pentacyclic distributions, which makes them unsuitable for correlation purposes. Finally, the great specicity of tricyclics and tetracyclics is demonstrated for the special case at hand. Thereby, it is possible to recognize that pure "A" oil presumed to be from a single source contains in fact a slight admixture of B components. Consequently, if careful detailed analysis of saturate polycyclics in considered for correlation purposes, it can facilitate the discovery of otherwise ignored phenomena.

REFERENCES

1. F AQUINO NETO, J.M. TRENDEL, A. RESTLE, J. CONNAN, P. ALBRECHT, in Advances in Organic Geochemistry, edited by M. BJOROY (Wiley, London 1981), pp. 659-667.
2. O.A. AREF'YEV, V.M. MAKUSHINA, A.A. PETROV, Int. Geol. Rev., 24, 6, 723-728 (1982)
3. F. BEHAR, R. PELET, J. Anal. Appl. Pyr., 7, 121-135 (1984)
4. BEHAR, R. PELET, J. ROUCACHE, Org. Geochem., 6, 587-595 (1984)
5. F. BEHAR, P. ALBRECHT, org. Geochem., 6, 597-604 (1984)
6. F. BEHAR, R. PELET, J. Anal. Appl. Pyr., 8, 173-187 (1985)

7. H. CASTEX, Rev. Inst. Fr. Pet. 40, 2, 169-189 (1985)

8. J. CONNAN in Advances in Petroleum Geochemistry, edited by J. BROOKS and D. WELTE (Academic Press 1984), pp. 299-335.

9. A. VAN DORSSELAER, Thèse de Doctorat-ès-Sciences, Université Louis Pasteur, Strasbourg, France (1975)

10 C.M. EKWEOZOR, O.P. STRAUZ, in Advances in Organic Geochemistry, edited by M. BJOROY (Wiley, London 1981), pp. 746-766

11 A ENSMINGER, Thèse de Doctorat-ès-Sciences, Université Louis Pasteur, Strasbourg, France (1977)

12. J. ESPITALIE, K. S. MAKADI, J. TRICHET, Org. Geochem. 6, 365-382 (1984)

13. C. GOURLAOUEN, Thèse de Docteur Ingénieur, Université Paris VI, Paris, France (1984)

14. A.Y. HUC, F. BEHAR, J.C. ROUSSEL, in First International Symposium on Characterization of Heavy Crude Oils and Petroleum Residues (Technip, France 1984), pp. 99-103

15. J.M. HUNT in Petroleum Geochemistry and Geology (San Francisco W.H. Freeman and Co 1979)

16. D.J. KREULEN, Chem. Weekblad, 31, 758-761 (1934)

17. D. LEYTHAEUSER, A.S. MACKENZIE, R.G. SCHAEFFER, F.J ALTEBAUMER, in Advances in Organic Geochemistry edited by M. BJOROY (Wiley, London 1981), pp. 136-146

18. D. LEYTHAEUSER, R.G. SHAEFFER, Org. Geochem., 6, pp. 671-681 (1984)

19. A.S. MACKENZIE, D. MCKENZIE, Geol. Mag., 120, 417-528 (1983)

20. J.C. MONIN, R. PELET in First International Symposium on characterization of Heavy Crude Oils and Petroleum Residues (Technip, France 1984) pp. 104-108

21. J. Ph PFEIFFER, R.N. SAAL, J. Phys. chem., 42, 139-149 (1940)

22. I. RUBINSTEIN, C. SPYCKERELLE, O.P. STRAUZ, Geochim. Cosmochim. Acta, 43, 1-6 (1979)

23. J RULLKOTTER, Z. AIZENSHTAT, B. SPIRO, Geochim. Cosmochim. Acta, 48, 151-157 (1984)

24. W.K. SEIFERT, Geochim. Cosmochim. Acta, 42, 473-484 (1978)

25. W.K. SEIFERT, J.M. MOLDOWAN, in Advances in Organic Geochemistry edited by A.G. DOUGLAS and J.R. MAXWELL (Pergamon Press, Oxford 1979) pp. 229-237

26. B.P. TISSOT, D.H. WELTE in Petroleum Formation and Occurance (Springer Verlag 1984)

27. M. VANDENBROUCKE in Advances in Organic Geochemistry edited by H.R. GAERTNER and H. WEHNER (Pergamon Press, Oxford 1971), pp. 547-565.

CORRELATION OF OIL-SOURCE BIOMARKERS IN RENQIU OILFIELD, NORTH CHINA

BAOQUAN LIU, XIANZHANG ZENG, SHUZHEN LIU
Research Institute of Petroleum Exploration
& Development, North China
Renqiu County, Hebei Province, China

Using biomarkers such as steranes and terpanes, geochemical correlation has been made for the crude oil
-source rocks of Renqiu oil field --- the biggest
"buried hill" oil pool in China. It is concluded
that Renqiu oil in Lower Paleozoic and Precambrian
marine carbonate rocks came from Lower Tertiary lacustrine source rocks in adjacent depressions along
the fault plane and unconformity interface, and it
is categorized as an oil pool of "Tertiary oil stored in an older reservoir." Furthermore, the geochemical
correlation of steranes and terpanes shows that some
seepages in the outcrop of Precambrian carbonate
rocks in North China are generated from the carbonate rocks themselves and they are primary seepages.

FOREWORD

Renqui oil field is located in the middle part of Jizhong
east depression area of North China(Fig.1). The carbonate
rock reservoirs of Middle and Upper Proterozoic, Cambrian
and Ordovician are the principle productive formations,
while some crude oil is produced from sandstones of
Lower Tertiary Oligocene Sha-he-jie formation and Dong-
ying formation. Up to date, Renqui oil field is the big-

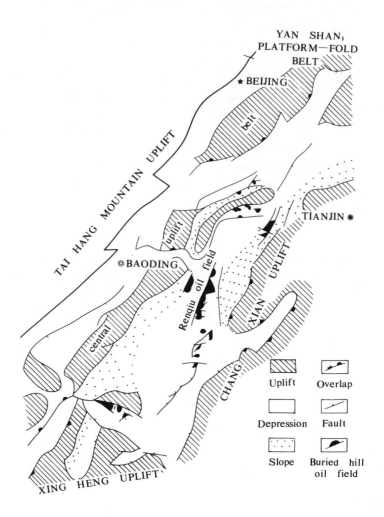

FIGURE 1   Location map of Jizhong Basin & Renqiu oil field

gest buried hill oil pool in China.

Its principle part consists of Wu-mi-shan formation siliceous dolomite of Middle and Upper Proterozoic with developed fissures and pores, becoming a fine reservoir.

Through a large fault with a throw of 3000m, its west side is in direct contact with dark mudstones of Lower

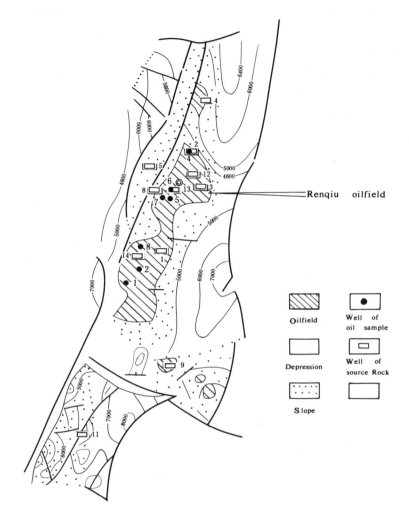

FIGURE   2   Distribution   of   samples   of   oil—source   Rocks   correlation

Tertiary Oligocene Es1 and Es3 formations; while through an
overlapping unconformity interface between Lower Tertiary
and Ordovician, Cambrian and Middle and Upper Proterozoic,
its east side and north and south ends are in indirect
contact with Es3 dark mudstones(Fig.2).

Jizhong depression is located at the west part of Bo-hai oil-bearing basin.. The Lower Tertiary terrestrial sedimentary area covers about 26,000 sq km with a thickness of 5,000-10,000m, making up the upper structure, and forming the principle oil-generating formation for the whole Bo-hai basin. The marine carbonate rocks are also well developed in Lower Paleozoic Ordovician, Cambrian and Middle and Upper Proterozoic of North China with a total thickness of 5,000-10,000m. In most of the Jizhong depression, they are directly overlapped by the Lower Tertiary, making up the lower structure of the depression, and emerging on the surface in the mountains both west and north of the depression. In the outcrop area of Yanshan folded belt, a total of 41 oil seepages have been found. Therefore, it is significant to reveal the formation of buried hill oil pools by making clear whether the crude oil of Renqiu buried hill oil pool is from the Lower Tertiary terrestrial source rocks or from the ancient marine carbonate rocks themselves. So by comparing biomarkers and stable carbon isotopes of crude oil, oil seepages and source rocks, we confirmed that the crude oil of Renqiu buried hill is mainly from the Lower Tertiary source rocks in the peripheral trough. We call this phenomenon "Tertiary oil stored in an older reservoir." The seepages in the peripheral outcrop area of the depression are mainly from the carbonate rocks themselves with a feature of "Older oil stored in an older reservoir".

CORRELATION OF OIL-SOURCE IN RENQIU OILFIELD
In Renqiu oil field, we collected 8 oil samples from 5 beds of Oligocene to Precambrianago, and 11 Oligocene terrestrial

TABLE 1 Formation and Depth of Samples of Oil—Source Rocks Correlation

Crude Oil Sample No.	Depth(m)	Formation	Source Rock Sample No.	Depth(m)	Formation	Lithology
1	2955	Ed	1	3460	Es3	Mudstone
2	2431.6	Ed	2	2841	Es1	Mudstone
3	2948	Es2	3	3308	Es3	Mudstone
4	3969	O	4	4399	Ek	Mudstone
5	3128	Є	5	3491	Es1	Mudstone
6	3311.4	Є	6	3438	Es3	Mudstone
7	2795	Zw	7	3722	Es3	Mudstone
8	3409	Zw	8	4280	Es3	Mudstone
			9	3376	Es1	Mudstone
			10	4006	Es3	Mudstone
			11	3920	Es1	Mudstone
			12	3505.38	O	Carbonate Rock
			13	3249.99	Є	Carbonate Rock
			14	3077.03	Zw	Carbonate Rock
9	Crop Seepage	Є	15	Crop	Є	Carbonate Rock
10	Crop Seepage	Zt	16	Crop	Zt	Carbonate Rock

TABLE 2   Physical - Chemical Property of Crude Oil in Renqiu Oilfield

Formation	Density d_4^{20}	Viscosity 50°C	Pour Point °C	Wax %	Sulfur %	Resin + Asphal-tine(%)	Ni Por-phyrin (ppm)	Trace Element (ppm)		
								Ni	V	V/Ni
Ed	0.8887	50.07	36	14.38	0.32	47.38	15.5	4.10	0.30	0.07
Es2	0.8793	72.44	40	18.26	0.31	30.70				
0	0.8792	41.34	37	17.70	0.25	20.90	2.0	0.40	0.10	0.25
ℯ	0.8788	60.18	37	24.53	0.13	25.57	13.55			
Zw	0.8798	35.21	33	22.90	0.26	14.90	26.43	0.50	0.10	0.20

source rock samples in the adjacent trough. In addition, we
collected 5 marine carbonate rock samples(3 hole and 2
outcrop samples) and another 2 seepages in the outcrop area
for comparison. The locations of sampling are shown in Fig
2. Table 1 summarizes the beds and depths of formations.

2.1. Physical and Chemical Properties of Renqiu Buried Hill Crude Oil

The physical and chemical properties of Renqiu buried
hill crude oil resemble those of the Lower Tertiary (Table
2). Both possess the common characteristics of terrestrial
crude oil such as high specific gravity, high viscosity,
high pour point, high wax content, low sulfur content and
low V/Ni ratio. This indicates that they probably have
similar sources. They are different from the crude oil of
low wax content (2.7%) and high sulfur content (7%)
in Palaeozoic marine limestone in China Sichuan basin,
also they are quite different from the seepages in carbonate rocks in the peripheral mountain area (low wax
content 0.6-2.2%).

2.2. Comparison with the Compositional Ternary Diagram of Three Steranes

Both Renqiu buried hill crude oil and most Lower Tertiary
source rocks are characterised by their high content of
$5\alpha-C_{27}$ steranes and low content of $5\alpha-C_{29}$ steranes. By
contrast, the buried hill and outcrop area carbonate rock
are characterised by their low content $5\alpha-C_{27}$ steranes
and high content of $5\alpha-C_{29}$ steranes. A practical method
to establish the correlation of crude oil and source rocks
is through a compositional ternary diagram of three steranes
of 5α configuration ($5\alpha-C_{27}$, $5\alpha-C_{28}$ and $5\alpha-C_{29}$ steranes).
Renqiu crude oil, adjacent source rocks and the

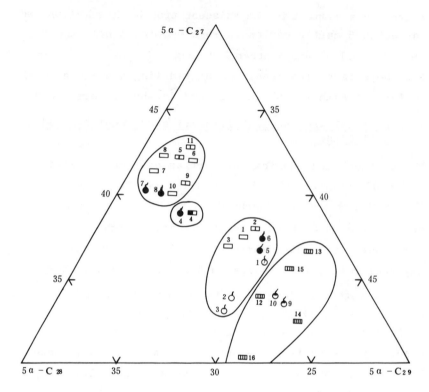

FIGURE 3   Triangle Diagram showing the composition of three
steranes of crude oil and seepages and source Rocks

♠—Buried hill Crude oil   ♂—Crop Seepage

♂—Tertiary Crude oil

▥—Es₁ Source Rock   ▭—Es₃ Source Rock

▰—E_K Source Rock   ▥—Carbonate

peripheral outcrop area seepages are all plotted into a
ternary diagram (Fig. 3), so that their genetic relation⁻
ships are shown clearly.

    As shown in Fig. 3: (1) spots of carbonate rocks and ¹
oil seepages are all at the lower right, farthest away
from the spots of crude oil of Wu-mi-shan formation ——
the principle productive formation in Renqiu ôil field,

indicating no genetic relationship among them; (2) spots of
crude oil of Renqiu Wu-mi-shan formation (Zw) are close to
that of source rocks in the peripheral trough, indicating
that they are from the source rocks of peripheral troughs
Es1 and Es3, and they belong to a mixed oil source; (3)
Cambrian and Ordovician (C-O) crude oils are close to the
nearest Lower Tertiary source rocks only. Their source is
from the nearby matrix with short distance migration; (4)
Lower Tertiary Oligocene crude oils are also from adjacent
Es1 and Es3 source rocks.

2.3. Composite Comparison with Sterane and Terpane Parameters

Fig. 4 is a comparison diagram for sterane and terpane
parameters and Pr/Ph of Renqiu oil and source rocks.

Obviously as shown in Fig. 5, after the crude oil is
expelled from Es3 oil-generative matrix, it migrates along
the fault and unconformity interface westward in a short
distance into the Ed and Es2 reservoir. Comparison of the
results of sterane and terpane parameters are in concord-
ance with the geological conditions.

In Renqiu oil field, Ordovician oil is different from
Wu-mi-shan (Zw) oil, and it contains apparent C_{28} Hopanes.
Various sterane and terpane parameters of Ordovician oil
are very similar to those of Kongdian formation (Ek) source
rocks in the north depression of the oil field (Fig. 4b).
Its Ni porphyrin content (2ppm) is less than that of the Zw
formation oil (26.43 ppm), but it closely approximates the
north depression oil ($<$5ppm). As shown in Fig. 6, after
being expelled from the north Ek source rocks, the crude
oil migrates along the unconformity surface southward and
accumulates in the Ordovician reservoir.

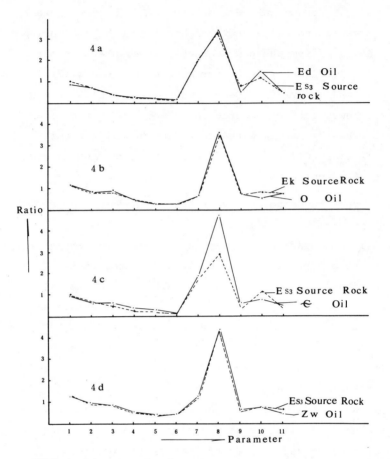

FIGURE 4　Diagram showing the oil to rock sterane and terpane correlation parameter in Renqiu oil field

$$1. \frac{5\alpha - C_{27}}{5\alpha - C_{29}} \quad 2. \frac{5\alpha - C_{28}}{5\alpha - C_{29}} \quad 3. \frac{5\beta - C_{27}}{5\alpha - C_{27}} \quad 4. \frac{20S - C_{29}}{(20S + 20R) \, C_{29}}$$

$$5. \frac{(5\beta + 14\beta) \, C_{29}}{\Sigma \, C_{29}} \quad 6. \frac{Diasterane}{5\alpha (C_{27} + C_{28} + C_{29})} \quad 7. \frac{Tm}{Ts} \quad 8. \frac{C_{30}Hop.}{(C_{29} + C_{30}) \, Mor.}$$

$$9. \frac{C_{29}Hop.}{C_{30}Hop.} \quad 10. \frac{Gammacerane}{C_{31}Hop. / 2} \quad 11. \frac{Pr.}{Ph.}$$

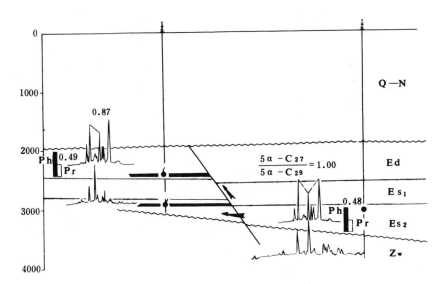

FIGURE 5   Cross section showing the oil—source correlation   of
             the Lower Tertiary oil pool in Renqiu   oil field

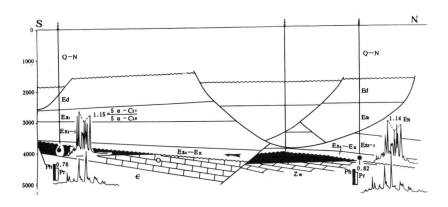

FIGURE 6   Cross section showing oil—source correlation   of Ordovi-
             cian oil pool in the north slope of Renqiu
             oil field

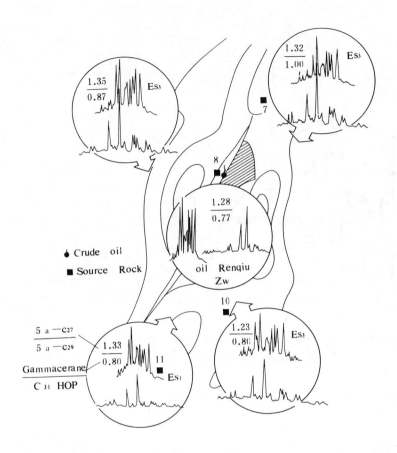

FIGURE 7   Plan view of the oil-source correlation of Wu-
mishan grou oil pool in Renqiu oil field

Renqiu Cambrian oil is similar to adjacent Es3 source
rocks in most parameters (Fig. 4c). The oil is mainly from
adjacent Es3 source rocks. However, parameters 8 and 9
show differences. Thus, the Cambrian pool is an in-place
oil pool with some additional contributing sources.

Figures 4d and 7 show the correlation of steranes and

terpanes between Renqiu crude oil of Middle and Upper Proterozoic Wu-mi-shan formations -- the principle productive
formations in the Renqiu buried hill oil field and source
rocks. It is shown that the crude oil sterane and terpane
parameters and mass chromatogram are similar to that of Es1
and Es3 source rocks in the peripheral trough, indicating
that the buried hill crude oil is from the source rocks of
the peripheral trough.

The oil-source correlation profile is shown in Fig. 8.
It indicates that the Renqiu Wu-mi-shan reservoir could be
in direct contact with Es1 and Es3 source rocks of the peripheral trough through a fault and unconformity interface.
The type of organic matter in the source rocks is fine
(type II, i.e. humic-sapropelic), their hydrocarbon content
is high (1008-1742 ppm), and the volume of oil-generated is
large. It is the best source rock in Jizhong region. The
crude oil expelled from the matrix could migrate along the
fault and unconformity interface and accumulate in the
buried hill reservoir. For just this reason, Renqiu oil
pool could become the largest buried hill oil field because
it is richly endowed by nature with a sufficient nearby oil
source, multiways of migration and fine reservoir conditions.

2.4. Comparison with Carbon Isotopes of Crude Oil Compo-
sition

Using the method of carbon isotope type curves suggested by
W.J. Stahl,[3] we make a comparison of five beds of oil in
the Renqiu oil field, their source rocks and outcrop seepages (Fig. 9) in order to verify the sterane and terpane
correlation results.

Fig. 9 shows that carbon isotopes of the various fractions of five crude oils from different beds in Renqiu oil

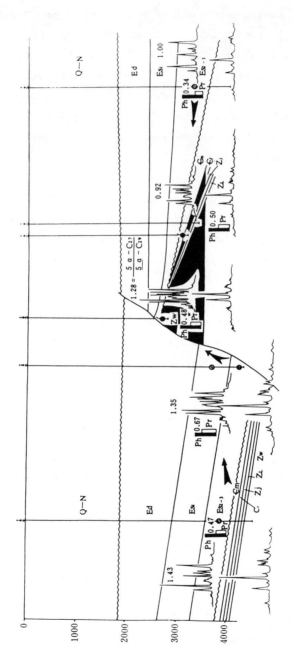

FIGURE 8   Cross section showing oil—source correlation of Buried hill oil pool in Renqiu oil field

field are close together. The kerogen of adjacent Es3 and

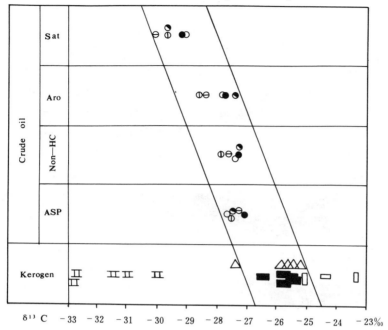

FIGURE 9   Carbon-isotope correlation diagram for frac-
             tions of Renqiu crude oil and keroqen of the
             source   rocks

field are close together. The kerogen of adjacent Es3 and
Es1 source rocks are all within the range of extension lines
of carbon isotope of crude oil fractions. By contrast,
kerogen of Es2 and Es4 - Ek source rocks and carbonate rocks
are outside of the extension lines, confirming that the five
crude oils in Renqiu oil field have similar sources. All
of them are from Es3 and Es1 source rocks. The ratio of
carbon isotopes of the carbonate rock kerogen is quite different from Renqiu crude oil, indicating that there is not

any genetic relationship between them.

2.5. Comparison with $5\alpha-C_{27}/5\alpha-C_{29}$ vs $\delta^{13}C\permil$ Diagram

Generally, the carbon isotope $\delta^{13}C\permil$ of various oil beds in Renqiu oil field is -28.22 — -29.12%. The $\delta^{13}C\permil$ of extract of Lower Tertiary source rocks is generally -26.80 — -28.55%, while the $\delta^{13}C\permil$ of carbonate rock chloroform extract and outcrop area seepages is as low as -30.70 — -33.00%. It is easy to distinguish them by plotting the sterane parameter $5\alpha-C_{27}/5\alpha-C_{29}$ as ordinate and $\delta^{13}C\permil$ as abscissa(Fig.10).

As shown in Fig.10, five beds of crude oil in Renqiu oil field and Lower Tertiary source rocks are in the same region (top right), while the carbonate rocks and the

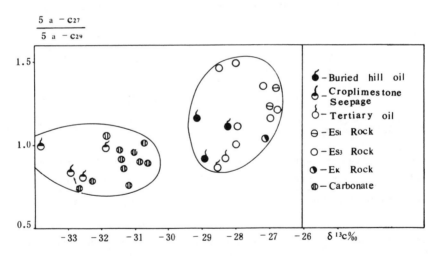

FIGURE 10   Correlation   Diagram   of $\dfrac{5\alpha-C_{27}}{5\alpha-C_{29}}$ sterane $-\delta^{13}c\permil$ for **Ren—**

qiu   Crude oil,   seepages   and   source   Rocks

seepages are in another region (lower left). Again it indicates that five beds of crude oil in Renqiu oil field are all from Es3 and Es1 lacustrine source rocks, with no apparent genetic relationship with marine carbonate rocks. Both Fig. 10 and Fig. 3 show that in the peripheral outcrop area, seepages of Ordovician, Cambrian and Middle and Upper Proterozoic are different from buried hill crude oil of Renqiu oil field. Their sterane parameters and carbon isotope ratios approximate close to that of the carbonate rock. This indicates that the seepages of outcrop area are mainly from the carobnate rocks themselves, with a feature of "older oil stored in an older reservoir".

CONCLUSION

1. Sterane and terpane parameters and carbon isotope $\delta^{13}C\%$ ratios of five crude oil beds in Renqiu buried hill oil field are similar to those of Lower Tertiary crude oils and lacustrine source rocks, with no genetic relationship with older marine carbonate rocks. The crude oils of Renqiu buried hill are mainly from Lower Tertiary Es1 and Es3 lacustrine source rocks in the peripheral trough, with a feature of "Tertiary oil stored in an older reservoir".

2. The biomarkers and carbon isotope of the seepages in the peripheral outcrop area are quite different from Renqiu buried hill crude oil, but somewhat similar to those of carbonate rocks, indicating that the seepages are principally from the carbonate rocks themselves.

REFERENCES

1. Zeng Xianzhang et al, Distribution Characteristics and Geological Significance of Biomarkers in Crude Oil and Source Rocks of Jizhong Depression, Abstract of 1st National Meeting on Organic Geochemistry, May 1982, pp 129-130
2. Zhang Zhencai et al, Oil & Gas Geology, Vol.4, No.1 pp 89-99 (1983)
3. W.J.Stahl, Geochim. Cosmochim. Acta, Vol.42, No.10, pp 1573-1577 (1978)

DETERMINATION OF THE MINERAL-FREE NICKEL AND VANADIUM CONTENTS OF GREEN RIVER OIL SHALE KEROGEN

GARY J. VAN BERKEL and ROYSTON H. FILBY
Department of Chemistry and Nuclear Radiation Center
Washington State University, Pullman, WA 99164-1300

In kerogen isolation a small amount of mineral impurity remains and analyses of kerogen for Ni and V must be corrected to a "mineral-free" basis. Green River oil shale was extracted with toluene and then demineralized with HCl and HCl-HF. Low temperature ashing (LTA) was used to isolate the acid-insoluble mineral impurity in an unaltered state while oxidizing the Ni and V complexes in the organic fraction to NiO and V_2O_5. Kerogen ash was treated with dilute HCl and isolated by filtration. Instrumental Neutron Activation Analysis (INAA) was used to measure the Ni and V concentrations in the kerogen, ash, and acid treated ash. The differences in Ni and V concentrations in the ash and acid treated ash, when corrected for pyrite oxidation, corresponds to the Ni and V content of the organic fraction. The correction to a mineral-free basis for V is small, while that for Ni is proportional to the pyrite content of the kerogen. Apart from V and Ni, only Co and Cr show significant organic association in the kerogen.

1. INTRODUCTION

Biomarker data present convincing evidence for the geochemical relationship of kerogen and petroleum. Research in this laboratory is aimed at understanding the kerogen/petroleum relationship by studying the geochemical evolution of trace element biomarker molecules, particularly Ni and V complexes, present in the kerogen. Nickel and vanadium are of particular interest because of the possible role of kerogen in the generation of Ni^{2+} and VO^{2+} porphyrin[1,2] and nonporphyrin[3,4]

complexes during catagenesis, and the overall role of kerogen in the geochemistry of Ni and V in organic-rich sediments.

Unfortunately, analysis of kerogen for Ni and V (and
other trace elements) is hampered by the intimate association of the organic and mineral matrices in sedimentary
deposits, and the insolubility of kerogen in organic solvents. Kerogen can be separated from its mineral matrix by
a variety of physical, chemical, and combined physical/-
chemical methods[5,6,7] but, regardless of the technique used,
the kerogen can never be isolated mineral-free. The major
residual mineral is usually pyrite (FeS_2) but zircon
($ZrSiO_4$), rutile (TiO_2), and a variety of neoformed fluorides can also be present. These minerals may contain
varying quantities of metal ions in isomorphic substitution.
Pyrite, for example, may contain significant quantities of
Ni and other elements such as As, Co, and Sb. The concentrations measured for those elements in the kerogen which
are associated with both the organic and mineral phases will
vary depending on the amount and type of residual mineral
matter. Therefore, trace element contents of kerogen must
be corrected to a "mineral-free" basis to represent their
organically bound fraction.

Correction for the major elements of the mineral component is usually accomplished using X-ray diffraction (XRD)
to identify the mineral phases, and high-temperature ashing
to quantify the mineral matter present. Such a correction
scheme requires a knowledge of the chemical reactions
involved in ash formation (e.g., $FeS_2 \rightarrow Fe_2O_3$)[5]. This correction method, for example, distinguishes between organic and
pyritic S based on the difference between total S in the
sample and pyritic S calculated by assuming all Fe in the

ash was present as FeS_2 in the sample.

The major drawback of the high-temperature ash correction method is its inability to distinguish between metal species present in the organic matrix and those associated with the mineral matrix. In the few studies published in which the metal contents of a kerogen have been measured[8-12], no correction for the trace element contribution of the mineral impurities was made.

In this paper, chemical/physical isolation of kerogen combined with low temperature ashing (LTA), acid treatment of the ash, and trace element analysis by instrumental neutron activation analysis (INAA) are used to distinguish between the organic and inorganic Ni and V contents of Green River oil shale kerogen. In addition, data for Co and Cr indicate that these elements are also bound, in part, to the kerogen matrix.

2.   EXPERIMENTAL

2.1 Samples and Sample Preparation

Mahogany-zone Green River oil shale (41 gal/ton, Union Oil, Calif.) was jaw crushed then ground to ~200 mesh using a ring mill. The shale (200 g) was batch extracted with 500 ml of toluene ($40^{o}C$) several times until a nearly color-less extract was obtained. The extracted shale was vacuum dried (~$80^{o}C$) and then demineralized using a procedure similar to that of Durand and Nicaise[7]. The kerogen was separated into float and sink fractions in the HCl-HF solution used for the demineralization, and each fraction was vacuum dried (~$80^{o}C$). Kerogens were batch extracted with toluene and methanol/acetone (1:9 v/v) to remove any bitumen (particularly Ni or V containing species) liberated

during the acid digestion until porphyrin absorbances at
410, 530, and 570 nm of the extract were not observed.
After extraction, the kerogen fractions were vacuum dried
($\sim 80^\circ$C). All extraction and demineralization steps were
carried out under N_2(g) to prevent oxidation of the kerogen.

Three Green River kerogen fractions (GRK-A, 3.35 wt%
ash; GRK-B, 11.70 wt% ash; GRK-C, 18.29 wt% ash (LTA)) were
used in this study.

2.2 Low Temperature Ashing

Kerogen samples were ashed using an LFE Model LTA-302 low
temperature asher. The asher was operated at 50 RF watts
and an O_2(g) flow rate of 2 mL/min. Samples of ~ 1 g of
kerogen were ashed in 5 cm diameter petri dishes and
samples were stirred every 2 hrs during the first 8 hrs.
Ashing was continued for an additional 16 hrs and the
samples were again stirred. Ashing was continued and
samples were weighed and stirred every 2 hrs until constant
weight was achieved (~ 35 hrs total ashing time).

2.3 Low Temperature Ashing Time Study

Asher operating conditions were the same as those above.
A 2.5 g sample of GRK-B was ashed for 36 hrs in a 10 cm
diameter petri dish. The sample was stirred hourly. The
ash product was sampled hourly up to 8 hrs and then at 10,
12, 14, 16, 20, 24, 28, and 36 hrs. The weight percent
sample remaining at each sampling time was calculated and
adjusted for each aliquot removed.

2.4 Chemical Treatment of Ash

Aliquots of each kerogen ash were weighed into 100 mL
beakers and 50 mL of 1M HCl added. A small quantity of
diluted Triton X100 was added to aid in wetting the ash.

Mixtures were stirred constantly and heated to 40°C for
1 hr. The leached ash was isolated by filtration (0.05 um
or 0.2 um Nucleopore polycarbonate filters), dried for
several hours at 100°C, and the weight percent ash recovered
was calculated. The ash samples from the time study were
treated similarly. Aliquots of the GRK-B and GRK-C ash
were treated in a similar manner with 1M NH_3.

2.5   Trace Element Analysis

Trace element concentrations were determined by INAA. Sample aliquots (10-100 mg) and appropriate standards were
weighed into clean 2/5 dram polyvials and reencapsulated in
2 dram polyvials, both of which were heat sealed. Samples
and standards were irradiated in the WSU TRIGA III reactor
and γ-ray spectra recorded using the Nuclear Data ND6700
Ge(Li) γ-ray spectrometer system. The nuclear reactions
and methods used were similar to those of Jacobs and
Filby[13].

2.6   X-Ray Diffraction Analysis

X-ray diffraction patterns of the shale, kerogen, ash, and
treated ash samples were obtained using a Norelco powder
diffractometer using Cu-K$_\alpha$ radiation. Samples were prepared as cavity or smear mounts. A Cameca Cambax electron
microprobe/scanning electron microscope with wavelength and
energy dispersive spectrometers was used to determine
qualitatively the distribution of various elements in the
kerogen, ash, and acid treated ash samples.

3.   RESULTS

Carbonates including calcite ($CaCO_3$), dolomite ($CaMg(CO_3)_2$),
and dawsonite ($NaAlCO_3(OH)_2$) were the major minerals iden-

tified by XRD in the shale. Silicate minerals detected
included quartz (SiO_2), albite ($NaAlSi_3O_8$), and analcite
($NaAlSi_2O_6 \cdot H_2O$). Pyrite (FeS_2) was also detected.

Pyrite was the major impurity in the three kerogen
fractions, although GRK-C contained a small amount of calcite and dolomite. The neoformed fluoride, ralsonite
($NaMgAl(F,OH)_6 \cdot nH_2O$), was detected in the ash of GRK-B
and GRK-C. The presence of calcite and dolomite after the
extensive acid demineralization procedure was unexpected.
The only mineral phase detected in the acid treated ash
samples was pyrite, because of the solubility of the calcite, dolomite, and ralstonite in dilute HCl. Organic
"skins"[5] may have protected the carbonates from acid attack
during the demineralization procedure, whereas the ralstonite
probably precipitated from the HCl-HF solution containing
high Al^{3+} and Na^+ concentrations.

Microprobe examination of the GRK-B treated ash showed
Ni X-ray intensities significantly greater than background
counts only when analyzing pyrite grains. The association
of Ni with the pyrite impurity confirms the necessity of
the "mineral-free" correction for determination of the true
organic Ni content of this kerogen. Other trace elements
found to be associated with the pyrite of the treated ash
include As, Co, Sb, and Se.

Other minor mineral phases were detected in the
treated ash. The most abundant of these contained only Ti
and was probably rutile (TiO_2) or some other TiO_2 polymorph.
Phases containing Al, Si, and varying quantities of Ca, K,
Mg, and Na were found, for the most part, on the surface of
pyrite grains. These phases may be precipitated fluorides
for which the pyrite provided nucleation sites.

Table 1 lists the concentration of 38 elements in kerogens GRK-A, B, and C. The concentrations for most elements show significant variation among the three fractions. The most abundant elements (i.e., Al, Ca, Fe, K, Na, and Ti) are consistent with the mineral impurities identified in the kerogen by XRD. The high concentration of Cl is probably due to residual HCl from the demineralization procedure.

Although ash content increases from GRK-A to GRK-C, there is no simple relationship between ash content and trace element content because the distributions of residual minerals are different in the three kerogen fractions. The increase in ash content of GRK-B over GRK-A results mainly from an increase in pyrite content whereas the high ash content of GRK-C is related to the occurrence of calcite, dolomite, and fluorides. The trace elements associated with these mineral phases are different, hence a direct relationship between the ash content of the kerogen and trace element concentrations is not observed. Fluctuation of the trace element content of the kerogen with varying amount and type of mineral impurity indicates at least partial association of the element with the mineral phase.

Nickel has a variable concentration in the three kerogen fractions which seems to be related to the pyrite content. The vanadium content, however, seems to be independent of the amount and type of mineral impurity present. These data indicate that V may be bound only in the kerogen matrix, whereas Ni is probably associated with both the mineral and kerogen matrices. Inspection of the data in Table 1 also indicate that most elements investigated fall into the dual, organic/inorganic association category, while only a few may be totally bound to the kerogen.

TABLE 1   Concentrations (ug/g) of 38 Elements in Kerogen
          GRK-A, GRK-B, and GRK-C.

Element	Elemental Concentrations X±s.d. (ug/g)*		
	GRK-A (3.35 wt% ash)	GRK-B (11.70 wt% ash)	GRK-C (18.29 wt% ash)
Al	1730 ± 60	1430 ± 150	10500 ± 230
As	51.6 ± 1.3	268 ± 9	213 ± 3
Ba	122 ± 11	70 ± 11	247 ± 15
Br	10.6 ± 1.3	5.22 ± 0.50	3.78 ± 0.45
Ca	2110 ± 80	800 ± 180	2480 ± 420
Ce	16.7 ± 3.1	11.4 ± 1.8	24.7 ± 0.3
Cl	N.M.	15300 ± 330	16600 ± 250
Co	12.2 ± 0.3	26.7 ± 3.2	62.3 ± 2.5
Cr	16.0 ± 1.2	26.4 ± 2.9	23.4 ± 1.2
Cs	0.75 ± 0.14	0.576 ± 0.060	1.70 ± 0.18
Eu	0.360 ± 0.037	0.129 ± 0.015	0.288 ± 0.031
Fe	6380 ± 39	38100 ± 4700	36400 ± 2100
Ga	15.6 ± 0.4	15.3 ± 0.1	14.9 ± 1.8
Hf	1.07 ± 0.14	2.02 ± 0.22	2.46 ± 0.11
Hg	3.82 ± 0.23	27.7 ± 3.4	2.72 ± 0.15
K	393 ± 40	605 ± 29	1400 ± 300
La	11.5 ± 0.3	8.19 ± 0.10	15.6 ± 0.5
Lu	0.098 ± 0.014	0.151 ± 0.009	0.158 ± 0.017
Mg	<500	<1000	1690 ± 240
Mn	N.M.	6.25 ± 0.05	7.93 ± 0.23
Na	486 ± 16	365 ± 5	7710 ± 230
Nd	6.94 ± 0.88	5.8 ± 1.8	9.7 ± 2.8
Ni	38.0 ± 1.5	79.7 ± 9.5	93 ± 10
Rb	8.4 ± 2.3	13.5 ± 1.4	23.7 ± 3.9
Sb	1.74 ± 0.15	4.59 ± 0.50	5.87 ± 0.15
Sc	0.717 ± 0.030	0.624 ± 0.096	1.346 ± 0.011
Se	5.63 ± 0.26	16.8 ± 2.0	14.80 ± 0.81
Sm	1.03 ± 0.08	0.653 ± 0.010	1.19 ± 0.03
Sr	54 ± 14	<50	90 ± 10
Ta	1.22 ± 0.05	1.51 ± 0.14	1.35 ± 0.04
Tb	0.098 ± 0.020	0.174 ± 0.039	0.198 ± 0.025
Th	2.75 ± 0.15	1.99 ± 0.26	3.65 ± 0.10
Ti	527 ± 13	626 ± 91	908 ± 26
U	3.39 ± 0.78	1.45 ± 0.49	3.06 ± 0.19
V	29.2 ± 1.3	31.2 ± 2.7	29.9 ± 1.8
Yb	0.82 ± 0.17	0.83 ± 0.10	1.06 ± 0.06
Zn	31.3 ± 2.0	68.0 ± 7.7	46.6 ± 0.4
Zr	61.5 ± 0.1	74 ± 36	68 ± 10

N.M. = not measured
<value = counts for element less than 2 times the std. dev.
         of background counts at photopeak energy
* X±s.d. = mean concentration ± std. dev. of 3 replicates

TABLE 2   Percent Retention of 38 Elements in the Low
Temperature Ash of Kerogens GRK-A, GRK-B, and
GRK-C.

Element	% Retention ± s.d.*		
	GRK-A (3.35 wt% ash)	GRK-B (11.70 wt% ash)	GRK-C (18.29 wt% ash)
Al	84.8 ± 3.1	96 ± 10	96.7 ± 3.1
As	89.3 ± 2.3	91.5 ± 3.7	138 ± 4
Ba	55.7 ± 6.3	85 ± 14	95.4 ± 6.2
Br	2.66 ± 0.80	9.9 ± 1.2	10.8 ± 1.5
Ca	76.6 ± 4.7	105 ± 26	103 ± 19
Ce	109 ± 20	93 ± 15	95.8 ± 41
Cl	n.d.	0.31 ± 0.01	0.77 ± 0.10
Co	98.3 ± 2.2	101 ± 12	54.2 ± 2.7
Cr	112 ± 14	102 ± 11	113 ± 10
Cs	68 ± 13	96 ± 10	95 ± 11
Eu	52.9 ± 5.8	88 ± 10	96 ± 11
Fe	100 ± 1	101 ± 13	117 ± 4
Ga	91.2 ± 3.2	100 ± 5	182 ± 23
Hf	93 ± 12	100 ± 11	137 ± 7
Hg	n.d.	9.67 ± 0.72	13.2 ± 6.0
K	82.5 ± 8.7	88.7 ± 5.3	91 ± 23
La	89.8 ± 2.3	90.6 ± 2.8	115 ± 4
Lu	114 ± 16	90.1 ± 5.9	117 ± 13
Mg	n.d.	n.d.	81 ± 25
Mn	n.d.	91.7 ± 0.9	95.9 ± 6.4
Na	86.1 ± 2.8	88.3 ± 2.9	111 ± 3
Nd	76 ± 28	101 ± 32	92 ± 29
Ni	102 ± 22	99 ± 13	102 ± 14
Rb	97 ± 27	101 ± 18	102 ± 24
Sb	101 ± 20	100 ± 11	109 ± 5
Sc	99.4 ± 4.2	100 ± 16	98.4 ± 9.0
Se	65.3 ± 4.6	82.9 ± 9.9	99.5 ± 7.1
Sm	91.8 ± 7.3	95.3 ± 3.9	108 ± 4
Sr	88 ± 32	n.d.	94 ± 12
Ta	110 ± 5	99.5 ± 9.3	94.4 ± 3.0
Tb	82 ± 23	83 ± 30	104 ± 27
Th	100 ± 11	99 ± 13	98 ± 11
Ti	95.6 ± 5.2	100 ± 18	105 ± 7
U	87 ± 21	80 ± 28	75.9 ± 7.3
V	92.6 ± 4.4	98.6 ± 9.0	97.9 ± 6.2
Yb	82 ± 17	94.3 ± 11	115 ± 8
Zn	98.0 ± 6.4	101 ± 12	90.1 ± 3.0
Zr	61.8 ± 2.2	86 ± 43	134 ± 32

* % Retention = $([X]_{ash} \cdot f_{ash} / [X]_K) \cdot 100 \pm s.d.$
(defined in text)

s.d. = propagated standard deviation for 3 replicate
analyses of kerogen and ash

n.d. = $[X]_K$ not measured

To determine the mineral-free trace element contents, the kerogen was low temperature ashed to constant weight and then leached with 1M HCl. The LTA procedure, which is used to isolate mineral matter from coal[14], oxidizes the organic matter to CO_2, H_2O, NO_x, SO_2, etc., and leaves the mineral matter relatively unaltered. Table 2 shows the effect of the ashing process on the elements in the kerogen. Except for the volatile elements Br, Cl and Hg, retention of most elements in the ash is not consistently different from 100% at the 99% confidence limits in all three kerogens.

Under the oxidizing conditions involving atomic oxygen, organically bound metal species are converted to their respective oxides and if the metal oxide is soluble in dilute acid it is possible to differentiate the fraction of the metal in the mineral matrix (insoluble in acid) from that in the kerogen matrix (soluble). Table 3 shows the percentage of 35 elements leached from the low temperature ash of the three kerogens by acid treatment, and Table 4 the percentage of the elements leached from the low tempera- ture ash of GRK-B and GRK-C by treatment with 1M NH_3. The similarities and differences between the amount of an ele- ment removed during the two treatments are related to the nature of the elemental species present in the ash.

The mineral-free kerogen trace element concentration is calculated from

$$[X]_{MFK} = ([X]_{ash} - [X]_{Lash} \cdot f_{Lash}) \cdot f_{ash} \qquad (1)$$

where $[X]_{MFK}$, $[X]_{ash}$, and $[X]_{Lash}$ are the concentration of element X in the mineral-free kerogen, low temperature ash, and acid leached ash, respectively. f_{Lash} and f_{ash} are the weight fraction leached ash recovered relative to the low temperature ash and the weight fraction ash recovered from low

TABLE 3   Percentages of 35 Elements Leached from GRK-A,
GRK-B, and GRK-C Low Temperature Ash by 1M HCl.

Element	% Removed $\pm$ s.d.*		
	GRK-A (28.16%)	GRK-B (67.26%)	GRK-C (50.25%)**
Al	98.7 $\pm$ 0.1	94.5 $\pm$ 0.3	98.4 $\pm$ 0.1
As	58.6 $\pm$ 0.4	26.1 $\pm$ 2.1	16.9 $\pm$ 3.0
Ba	91.7 $\pm$ 3.1	80.9 $\pm$ 5.4	91.3 $\pm$ 1.9
Ca	--	--	--
Ce	85.4 $\pm$ 0.4	83.0 $\pm$ 0.5	70.4 $\pm$ 9.0
Co	92.2 $\pm$ 0.1	77.0 $\pm$ 2.5	70.2 $\pm$ 1.1
Cr	67.6 $\pm$ 3.3	65.6 $\pm$ 0.7	58.5 $\pm$ 3.2
Cs	--	87.8 $\pm$ 1.4	--
Eu	89.2 $\pm$ 1.3	76.7 $\pm$ 1.2	79.5 $\pm$ 1.4
Fe	54.0 $\pm$ 0.1	19.0 $\pm$ 2.6	14.1 $\pm$ 4.5
Ga	99.1 $\pm$ 0.1	62.4 $\pm$ 2.3	37.8 $\pm$ 3.0
Hf	39 $\pm$ 11	22.1 $\pm$ 2.3	19.9 $\pm$ 2.5
K	52 $\pm$ 11	--	--
La	77.8 $\pm$ 0.2	78.2 $\pm$ 0.7	73.7 $\pm$ 0.6
Lu	63.6 $\pm$ 1.4	55.6 $\pm$ 4.0	48.5 $\pm$ 2.6
Mg	--	--	--
Mn	N.M.	26.7 $\pm$ 0.5	17.5 $\pm$ 5.0
Na	99.4 $\pm$ 0.1	89.1 $\pm$ 0.4	99.5 $\pm$ 0.1
Nd	81.9 $\pm$ 4.6	--	65 $\pm$ 10
Ni	93.7 $\pm$ 2.3	66.4 $\pm$ 2.4	57.4 $\pm$ 3.6
Rb	--	53 $\pm$ 20	72.6 $\pm$ 5.8
Sb	78.6 $\pm$ 1.8	32.2 $\pm$ 1.5	35.9 $\pm$ 3.9
Sc	78.5 $\pm$ 0.2	62.6 $\pm$ 2.9	66.7 $\pm$ 3.9
Se	70.7 $\pm$ 2.2	23.8 $\pm$ 1.7	30.0 $\pm$ 4.5
Sm	90 $\pm$ 0.1	83.3 $\pm$ 1.9	80.0 $\pm$ 0.7
Sr	--	--	41.9 $\pm$ 6.4
Ta	84.5 $\pm$ 1.4	69.5 $\pm$ 0.3	79.9 $\pm$ 0.4
Tb	76.6 $\pm$ 5.4	34 $\pm$ 21	40 $\pm$ 16
Th	88.1 $\pm$ 1.2	81.2 $\pm$ 0.8	90 $\pm$ 11
Ti	22.8 $\pm$ 5.8	34.0 $\pm$ 7.7	$\leq$0
U	84.4 $\pm$ 2.0	79.0 $\pm$ 2.5	72.3 $\pm$ 2.8
V	97.6 $\pm$ 0.2	95.1 $\pm$ 0.3	92.0 $\pm$ 0.2
Yb	68.0 $\pm$ 3.5	46.3 $\pm$ 2.5	57.0 $\pm$ 5.0
Zn	--	95.3 $\pm$ 0.3	--
Zr	64.6 $\pm$ 5.6	39.4 $\pm$ 8.1	29 $\pm$ 14

* % Removed = 100 $\cdot$ [1.0 - ($[X]_{Lash} \cdot f_{Lash}/[X]_{ash}$)] $\pm$ s.d.
(defined in text)
**Percent recovery of leached ash relative to kerogen ash
in parentheses
s.d. = propagated standard deviation
-- = concentration remaining in ash below detection limit
N.M. = not measured

TABLE 4   Percentages of 35 Elements Leached from GRK-B
and GRK-C Low Temperature Ash by 1M NH_3 Treatment.

Element	% Removed ± s.d.*	
	GRK-B (97.18%)	GRK-C (64.4%)**
Al	62.2 ± 1.8	70.2 ± 0.8
As	≤0	≤0
Ba	0.7 ± 6.4	5.6 ± 3.8
Ca	≤0	23 ± 11
Ce	≤0	≤0
Co	≤0	≤0
Cr	6.4 ± 1.3	10.4 ± 7.7
Cs	94.1 ± 1.6	--
Eu	≤0	1.4 ± 4.8
Fe	≤0	0.3 ± 4.4
Ga	57.1 ± 2.9	66.4 ± 1.5
Hf	≤0	≤0
K	91.9 ± 1.4	92.2 ± 2.2
La	≤0	≤0
Lu	1.6 ± 5.0	≤0
Mg	29 ± 36	34 ± 18
Mn	≤0	N.M.
Na	85.7 ± 0.4	99.0 ± 0.2
Nd	29.4 ± 9.0	≤0
Ni	25.4 ± 4.5	11 ± 22
Rb	--	79 ± 21
Sb	≤0	18.3 ± 3.7
Sc	≤0	≤0
Se	16.8 ± 2.4	19.6 ± 4.2
Sm	≤0	_0
Sr	--	7 ± 15
Ta	≤0	0.1 ± 3.5
Tb	≤0	≤0
Th	≤0	≤0
Ti	≤0	≤0
U	≤0	≤0
V	≤0	48.4 ± 2.5
Yb	24.9 ± 3.2	9 ± 12
Zn	37.4 ± 2.0	41.1 ± 5.6
Zr	12 ± 13	≤0

* % Removed = 100 · [1.0 - ($[X]_{Lash}$ · f_{Lash}/$[X]_{ash}$)] ± s.d.
          (defined in text)
**Percent recovery of leached ash relative to kerogen ash
 in parentheses
s.d. = propagated standard deviation
--   = concentration remaining in ash below detection level
N.M. = not measured

temperature ashing of the kerogen.

Unfortunately, leaching of a metal ion from the ash may
not be solely the result of dissolution of oxides originating from the organically bound element in the kerogen. The
solubility of the Al and Na species in dilute acid and base
can be explained by the dissolution of ralsonite which pre-
cipitates with kerogen during the HCl-HF demineralizing
step. Other elements including Ba, Ca, Cs, Ga, Th, Zr, and
the rare earths which form stable fluorides may coprecipi-
tate with ralsonite and total or partial removal of these
elements results from the solubility of the fluorides in
dilute acid. The amount of these fluoride-forming elements
bound in the kerogen cannot be calculated using the present
method.

A significant quantity of Fe is removed from the ash
by acid treatments but essentially all of the Fe is retained
in the ash during treatment with base, consistent with the
presence of Fe_2O_3. This would indicate that either a large
quantity of Fe is organically bound in the kerogen (as much
as 7000 ug/g) or a portion of the FeS_2 in the kerogen is
oxidized to Fe_2O_3 during the ashing process, which is
probable because oxidation has been observed in the isolation of a mineral matter from coal[14]. That oxidation of
pyrite occurred during LTA is confirmed by the parallel
behavior of As and Fe during the acid and base treatments,
and by the association of As with the pyrite as shown by
microprobe examination. This behavior indicates that As is
oxidized to As_2O_3 in that fraction of the pyrite that has
been oxidized. Thus other trace elements associated with
the pyrite are partially leached by acid from the ash as a
result of pyrite oxidation and not because of an organic
association of the elements in the kerogen. From the micro-

probe data and from their common association with pyrite,
we can assume that, in addition to As, a portion of the Co,
Ni, Sb, and Se removed from the ash by acid treatment
results from pyrite oxidation. The mineral-free kerogen
concentration corrected for loss of an element from the
ash due to pyrite oxidation can be calculated using

$$[X]_{MFK'} = [([X]_{ash} - [X]_{Lash} \cdot f_{Lash}) - ([Fe]_{ash} -$$

$$[Fe]_{Lash} \cdot f_{Lash}) ([X]_{Lash}/[Fe]_{Lash})] \quad (2)$$

$$\cdot f_{ash}$$

where $[X]_{MFK'}$ is the concentration of element X in the
mineral-free kerogen corrected for loss of X during acid
treatment due to pyrite oxidation. $[Fe]_{ash}$ and $[Fe]_{Lash}$
are the concentration of Fe in the low temperature ash and
leached ash, respectively. This calculation assumes that
all Fe leached from the ash is due to pyrite oxidation
(i.e., $[Fe]_{MFK'} = 0$), all Fe in the treated ash is present
as pyrite, and all X in the treated ash is associated with
the pyrite. Further, it is assumed that the X/Fe ratio in
the pyrite of the treated ash is the same as the ratio of
X/Fe in the pyrite present in the kerogen.

Acid treatment removes V almost completely from each
of the three kerogen ashes. In 1M NH_3, leaching is incom-
plete, but complete removal has been shown to occur using a
strong base, 1M LiOH (GRK-C ash, Van Berkel, unpublished).
The reason for the difference in %V leached between GRK-B
and GRK-C by 1M NH_3 is not known. The behavior of the V in
the ash during acid and base treatments is consistent with
the chemistry of amphoteric V_2O_5, and indicates almost total
organic association of V in GRK-C. In contrast, $\leq 34\%$ of the
Ti is leached from the ash by acid treatment and none is

removed by base. This behavior suggests that the major
form of Ti in the kerogen is an acid resistant mineral, probably rutile (or other TiO_2 polymorph) or, less likely,
ilmenite ($FeTiO_3$). The elements Cr, Mn, U, and Zn are not
completely removed from the ash but their mineral component
is not clear. Traces of each element may be associated with
the fluorides or the pyrite. Chromium (as well as V) is a
known contaminant of rutile and Zr may be present as zircon.

4. DISCUSSION

4.1 Distribution of Trace Elements in Kerogen

Table 5 presents the kerogen ($[X]_K$), mineral-free kerogen
($[X]_{MFK}$), and pyrite loss corrected mineral-free kerogen
($[X]_{MFK'}$) trace element contents of the three kerogen fractions. Table 5 includes the mineral-free corrected concentrations for Al to illustrate the application of the correction to an element present in mineral form but leached from
the ash by acid treatment. No consistent mineral-free value
is obtained for such elements. For Al the mineral-free
value obtained is dependent on the Al content of the isolated kerogen. Similar results are obtained for all
elements present as soluble fluorides. The $[Fe]_{MFK}$ value
is dependent on the extent of oxidation of the pyrite during
ashing. The oxidation of pyrite also affects the mineral-
free concentrations calculated for the elements associated
with the pyrite (e.g., As, Co, Ni, Sb and Se).

The correction to a mineral-free basis for V is small.
In each kerogen fraction over 90% of the total V measured
in the isolated kerogen is organically bound. The $[Ni]_{MFK}$
values for the three fractions show better correspondence
than do the $[Ni]_K$ values. GRK-B and GRK-C have higher

TABLE 5   Concentrations (ug/g) of Seven Elements in Kerogen,
Mineral-Free Kerogen, and Mineral-Free Kerogen
Corrected for Pyrite Oxidation.

Element	Category*	Elemental Concentrations X±s.d. (ug/g)**		
		GRK-A (3.35 wt% ash)	GRK-B (11.70 wt% ash)	GRK-C (18.29 wt% ash)
Al	$[Al]_K$	1730 ± 60	1430 ± 150	10500 ± 200
	$[Al]_{MFK}$	1450 ± 20	1310 ± 20	9990 ± 230
	$[Al]_{MFK'}$	–	–	–
As	$[As]_K$	51.6 ± 1.3	268 ± 9	28 ± 3
	$[As]_{MFK}$	27.0 ± 0.3	58.7 ± 6.7	49.3 ± 9.8
	$[As]_{MFK'}$	4.64 ± 1.00	≤0	9.3 ± 15
Co	$[Co]_K$	12.2 ± 2.7	26.7 ± 3.2	62.3 ± 2.5
	$[Co]_{MFK}$	11.1 ± 0.9	20.6 ± 0.4	23.7 ± 10
	$[Co]_{MFK'}$	10.0 ± 0.9	14.2 ± 0.6	22 ± 10
Cr	$[Cr]_K$	16.0 ± 1.2	26.4 ± 2.9	23.4 ± 1.2
	$[Cr]_{MFK}$	12.1 ± 1.8	17.4 ± 0.1	15.4 ± 2.0
	$[Cr]_{MFK'}$	–	–	–
Fe	$[Fe]_K$	6380 ± 40	38100 ± 4700	36400 ± 200
	$[Fe]_{MFK}$	3450 ± 300	6380 ± 1050	6020 ± 1580
	$[Fe]_{MFK'}$	0	0	0
Ni	$[Ni]_K$	38.0 ± 1.5	79.7 ± 9.5	93.0 ± 10.0
	$[Ni]_{MFK}$	36.3 ± 8.1	52.1 ± 4.6	54.6 ± 8.1
	$[Ni]_{MFK'}$	33.4 ± 8.1	46.9 ± 4.7	48.0 ± 8.3
Sb	$[Sb]_K$	1.74 ± 0.15	4.59 ± 0.50	5.87 ± 0.15
	$[Sb]_{MFK}$	1.28 ± 0.09	1.39 ± 0.07	2.30 ± 0.31
	$[Sb]_{MFK'}$	0.407 ± 0.007	≤0	0.68 ± 0.18
Se	$[Se]_K$	5.63 ± 0.26	16.8 ± 2.0	14.8 ± 0.8
	$[Se]_{MFK}$	2.60 ± 0.20	3.00 ± 0.26	4.42 ± 0.82
	$[Se]_{MFK'}$	1.34 ± 0.21	≤0	2.72 ± 0.82
V	$[V]_K$	29.2 ± 1.3	31.2 ± 2.7	29.9 ± 1.8
	$[V]_{MFK}$	26.4 ± 0.4	29.2 ± 0.9	26.9 ± 0.6
	$[V]_{MFK'}$	–	–	–

*$[X]_K$, $[X]_{MFK}$, $[X]_{MFK'}$ = concentration of element in kerogen,
kerogen corrected for mineral content, and
kerogen corrected for mineral content and
pyrite oxidation, respectively

**X±s.d. = mean concentration ± propagated standard deviation

[Ni]$_{MFK}$ values than GRK-A because pyrite oxidation results
in an underestimation of the mineral Ni content. When the
loss of Ni from the ash due to pyrite oxidation is corrected
for (i.e., [Ni]$_{MFK'}$), the Ni concentrations of the three
kerogen fractions are not significantly different at the 95%
confidence level. Of the other elements investigated in
this study for which the [X]$_{MFK}$ or [X]$_{MFK'}$ values can be
calculated, only Co and Cr show a significant organic association. The low and variable concentrations calculated
for As, Sb, and Se do not rule out organic association but
show they do not play a significant role in the organic
matrix.

Additional information on the distribution of the elements is provided by the ashing time study. Plotting the
percentage of an element leached from the kerogen low
temperature ash by acid treatment versus ashing time (and
percent organic matter oxidized) results in distinctive
curves for organic, organic/inorganic, and inorganic association of the elements. The only elements which do not
follow these patterns are those present as fluorides.
Figure 1 compares the behavior of V (organic) and Ti
(inorganic). The %V leached from the partially ashed
kerogen increases with ashing time and is therefore
directly proportional to the amount of organic matter
oxidized. When oxidation of the organic matter is complete,
the %V leached remains at a constant level corresponding to
the organic V content of the sample. As can be seen, only
a very small fraction of the V is bound in an insoluble
mineral species. Aluminum (and also Na) shows behavior
which parallels that of V, but this results from the solubility of ralstonite in dilute acid.

The %Ti leached from the ash is not significantly

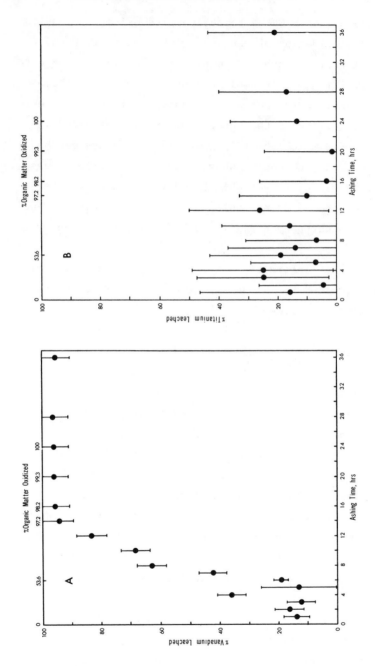

FIGURE 1   Percent Vanadium (A) and Titanium (B) Leached from GRK-B Low Temperature Ash by 1M HCl Versus Ashing Time and Percent Organic Matter Oxidized.

greater than zero over the entire ashing time. This confirms the presence of Ti as an HF-insoluble TiO_2 polymorph,
probably rutile. Similar curves are expected for an element
associated totally with an HF-insoluble mineral phase.

Chromium and nickel (Figure 2) show behavior intermediate
between that shown for V and Ti indicating that both elements are partially bound in both the kerogen and mineral
matrix. The %Cr leached from the ash increases with ashing
time indicating an organic association, but a large fraction
of the total Cr remains in the ash after complete oxidation
indicating mineral association. Nickel shows very similar
behavior except that a smaller percentage of the total Ni is
present in the mineral component.

Figure 3 reveals that the %Fe leached from the ash
reaches a constant level after only a few hours. This
behavior appears to be due to oxidation of a finely divided
component of the pyrite in the kerogen. There is a slight
increase in the %Fe leached from the ash after 100% of the
organic matter has been oxidized indicating further pyrite
oxidation, probably of larger mineral grains. The parallel
behavior between Fe and As (Figure 3) supports the oxidation
of pyrite as the explanation for the Fe behavior.

The oxidation of pyrite accounts for the large %Ni
leached from the ash early in the ashing process. Nickel
associated with the oxidized pyrite is converted to NiO and
is leached from the ash by dilute HCl. This effect combined with the removal of Ni originally organically bound
gives rise to the relationship seen in Figure 2. If the
%Ni leached from the ash due to pyrite oxidation is subtracted from the overall Ni curve, the direct relationship
between the %Ni leached and organic matter oxidized is much
more pronounced indicating organic combination of the

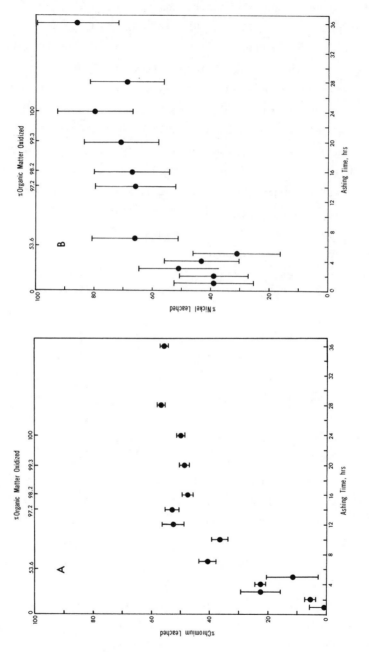

FIGURE 2   Percent Chromium (A) and Nickel (B) Leached from GRK-B Low Temperature Ash by
1M HCl Versus Ashing Time and Percent Organic Matter Oxidized.

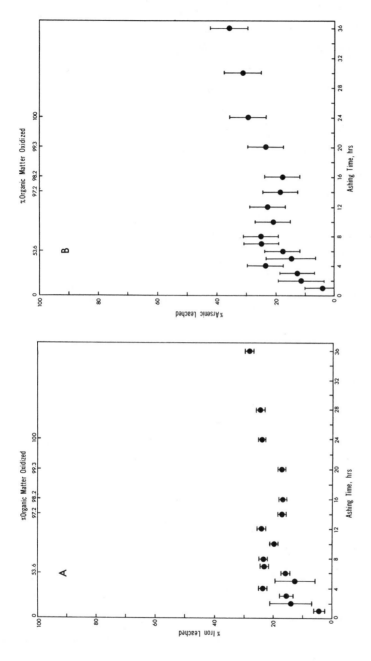

FIGURE 3   Percent Iron (A) and Arsenic (B) Leached from GRK–B Low Temperature Ash by 1M HCl Versus Ashing Time and Percent Organic Matter Oxidized.

element.

Figure 4 shows the elemental associations determined
for the Green River kerogen. These conclusions are based on
XRD and microprobe observations and on data from the acid
leaching of the kerogen ash. Of the elements studied, only
V, Ni, Cr, and Co appear to show any organic association.
Only V appears to be bound totally in the kerogen. Therefore, trace element analyses of kerogen prepared by HCl-HF
demineralization are probably accurate only for V while the
contents of all other elements must be corrected to a
mineral-free basis (as shown in Table 5). The extent of
the correction will be dependent on the amount and type of
mineral impurity remaining with the kerogen. It can be
concluded that the most accurate uncorrected results are
obtained if the kerogen is isolated with minimal mineral
impurity containing essentially only pyrite. As would be
expected, the smallest differences between the $[X]_K$ and
$[X]_{MFK}$ values were obtained for the purest kerogen fraction,
GRK-A. The presence of neoformed fluorides or other acid
soluble mineral impurities in the kerogen makes application
of this correction method to the elements involved in such
species impossible. This severely limits the number of
elements to which the mineral-free correction can be applied. Despite these limitations, this correction method
is capable of determining the true contents of Ni, V, and
a number of other elements in kerogens.

4.2 Geochemical Implications

The presence of Ni and V (and other trace elements) in the
kerogen is not unexpected. A number of studies present indirect evidence for the existence of organically bound metal species
in kerogen. Polar organics (e.g. asphaltenes) in many cases contain high concentrations of organically combined Ni, V, and

Trace Elements in Kerogen + Mineral Impurity			Elements
Organically Bound	Totally Organic		V
	Partially Organic		Ni, Co, Cr
Inorganically Bound	Residual Minerals (pyrite, rutile, zircon)	Major Components	Fe, Ti
		Minor Components	As, Co, Cr, Ni, Sb, Se, Zn, Zr
	Other Minerals (neoformed fluorides, calcite, dolomite, etc.)		Al, Ba, Ca, Cs, Ga, K, Mg, Mn, Na, Sr, Th, U, Zn, rare earths

FIGURE 4   Distribution of Organically-Bound and Inorganically-Bound Trace Elements in Green River Kerogen Containing Mineral Impurities.

other trace elements[13,15]. When these metals are incorporated into the asphaltene matrix is uncertain, but a
significant portion may have been bound in the kerogen
precursor.

Some evidence exists for a link between kerogen and
metalloporphyrins found in bitumens and crude oils. High
molecular weight porphyrin type complexes which have been
identified[3,4] are suggested to result from the thermal
breakdown of kerogen within which a tetrapyrrole complex
is bound.

Recently, Mackenzie et al.[1] examined the concentration
and type of VO^{2+} porphyrins isolated from the bitumens of a
suite of Toarcian shales from the Paris Basin. The data
show an expected decrease in DPEP/etio porphyrin ratio with
depth (thermal conversion) until approximately 2100m to
2450m. At this point, the total concentration of porphyrins
in the samples, average carbon number of the porphyrins, and
DPEP/etio ratio increase. Generation of VO^{2+} DPEP porphyrins with extended alkyl chains from kerogen was hypothe-
sized to account for these observations.

Baker and Louda[2] have found Ni^{2+} porphyrins in all but
the most recent sediments, while VO^{2+} porphyrins are more
commonly isolated from sediments which have undergone a
moderate degree of thermal stress. The authors conclude
that Ni^{2+} porphyrins are formed and exist in the sediment
primarily as free, or solvent extractable species, whereas
VO^{2+} porphyrins are believed to form in a bound, or non-
extractable state, linked to kerogen. When the imposed
thermal stress breaks the kerogen-porphyrin linkage, the
VO^{2+} porphyrins are released to the bitumen. At the present
time, no satisfactory explanation for the apparent kerogen
enhanced chelation and association of only VO^{2+} porphyrins

exists. In fact, as shown in this study, both Ni and V are present in the Green River kerogen.

Nickel and vanadium are likely bound in the kerogen matrix as stable tetrapyrrole structures; however, the mode of chelation of the other organically bound elements is not known. Cobalt and chromium are known to be bound in humic type substances but the stability of such complexes under HCl-HF treatments is questionable. It may be the case that other elements are actually bound in the kerogen but the severe acid demineralization treatment results in demetalla- tion for all but those elements bound to tetrapyrrole structures and those nonporphyrinic metals chelated in high concentrations.

The question of incorporation of these metals into the kerogen structure still needs to be addressed. It is not certain whether porphyrin species are bound to the kerogen or are trapped in a molecular sieve type network, or strongly adsorbed. In addition, the time and mechanism of incorporation of these metals or metal chelates into the kerogen matrix is unknown.

REFERENCES

1. A.S. MACKENZIE, J.M.E. QUIRKE, and J.R. MAXWELL, in Advances in Organic Geochemistry, 1979, edited by A.G. Douglas and J.R. Maxwell (Pergamon Press, Oxford, 1980), pp. 239-248.
2. E.W. BAKER and J.W. LOUDA, in Advances in Organic Geochemistry, 1981, edited by M. Bjorøy (John Wiley & Sons, Ltd., 1983), pp. 401-421.
3. M. BLUMER and W.D. SNYDER, Chem. Geol., 2, 35 (1967).
4. M. BLUMER and M. RUDMAN, J. Inst. Petrol., 56 (548), 99 (1970).
5. J.D. SAXBY, Chem. Geol., 6, 173 (1970).
6. J.D. SAXBY, in Oil Shale, edited by T.F. Yen and G.V. Chilingarian (Elsevier, Amsterdam, 1976), Chap. 6, pp. 103-128.

7.   B. DURAND and G. NICAISE, in Kerogen: Insoluble
     Organic Matter from Sedimentary Rocks, edited by
     B. Durand (Editions Technip, Paris, 1980), Chap. 2,
     pp. 35-53.
8.   K.W. RILEY and J.D. SAXBY, Chem. Geol., 37, 265 (1982).
9.   R.A. NADKARNI, in Geochemistry and Chemistry of Oil
     Shales, edited by F.P. Miknis and J.F. McKay (Ameri-
     can Chemical Society, Washington, D.C., 1983),
     Chap. 27, pp. 477-492.
10. K.M. JEONG and T.P. KOBYLINSKI, in Geochemistry and
     Chemistry of Oil Shales, edited by F.P. Miknis and
     J.F. McKay (American Chemical Society, Washington,
     D.C., 1983), Chap. 28, pp. 493-512.
11. B. SPIRO, D. DINUR, and Z. AIZENSHTAT, Chem. Geol.,
     39, 189 (1983).
12. A. SAOIABI, M. FERHAT, J.M. BARBEE, and R. GUILARD,
     Fuel, 62, 963 (1983).
13. F.S. JACOBS and R.H. FILBY, Anal. Chem. 55, 74
     (1983).
14. R.N. MILLER, R.F. YARZAB, and P.H. GIVEN, Fuel, 58,
     4 (1979).
15. B. HITCHON and R.H. FILBY, AAPG Bull., 68, 838 (1984).

ISOLATION, IDENTIFICATION AND CORRELATION OF PORPHY-
RINS FROM SOME CONTINENTAL CRUDE OILS, SOURCE ROCKS,
OIL SHALES AND COALS OF CHINA

ZHIQIONG YANG, ZHONGDI CHENG
Scientific Research Institute of Petroleum
Exploration and Development, Beijing
P.O. Box 910, Beijing
The People's Republic of China

The methodology of solvent extraction, TLC, HPLC, UV-
visible and fluorescence spectrophotometry applicable
to the analysis of porphyrins in the crude oils and
source rocks of Shengli, Liaohe, Renqui and Jianghen oil
fields, as well as the oil shale and coal of Fushum
and Maoming basins are described. Correlation of these
data and differences among them suggest the following:
1. Free porphyrin types based on electronic absorption
    spectrophotometry for crude oils and source rocks
    are of a mixed type; for oil shales, DPEP; for
    coals, ETIO.
2. Ratios of DPEP/ETIO for crude oils and source rocks
    are 2.2-3.4; for oil shales, 4.5; for coals, 0.3.
3. Maximum mass abundance of porphyrin molecules for
    crude oils and source rocks is m/z 462; for oil
    shales, m/z 476; for coals, m/z 478.
4. Concentration of metal porphyrins for crude oils
    are 53-424 ppm; for oil shales, 2.2-31 ppm; for
    coals, 0.25-2.0 ppm.

1. INTRODUCTION

The porphyrin compounds of continental crude oils and source
rocks of China have been studied[1, 2, 3] but few thorough

studies on their type have been made and the porphyrins of
oil shale and coal have been studied less frequently. In
the field of organic geochemistry it is worthwhile to study
the porphyrins of crude oils, source rocks, oil shales and
coals and discuss their general and special character, such
as the study of diagenesis, the maturity of the organic matter and the geochemical environment. Porphyrin compounds of
various samples mentioned above are analysed and some corre-
lative problems are discussed in this study.

2. EXPERIMENTAL

2.1. Samples and Their Analysis

All samples collected were of lower Tertiary (Eocene) age,
down to a depth of 2888m. A description of samples is sum-
arized in Table 1.

The methods used for separation and identification were
column chromatography, thin layer chromatography, HPLC, MS,
visible and fluorescence spectrophotometry. The porphyrins
of all samples (except No. 8 and No. 10) were treated for
demetallation and the free porphyrins were identified by LC,
MS and visible spectrophotometry. The free porphyrins of
sample No. 5 were separated by thin-layer chromatography and
its different components were identified by LC, MS, visible
and fluorescence spectrophotometry.

TABLE 1   Description of samples

Sample type	Area	Age	Depth(m)	Note
No.				
1. Crude oil	Well Yi 20 Shengli	Lower	2885–2888	
2.	Well 104 Renqiu	Tertiary	2404–2405	
3.	Qiangiang Jianghen		Mix	
4.	Gaosheng Liaohe		Mix	
5. Oil Shale	Fushum	Lower	Outcrop	F–2
6.	Fushum	Tertiary		F–3
7.	Fushum			F–5
8.	Maoming			M–1
9. Coal	Fushum	Lower	Outcrop	F–4
10.	Maoming	Tertiary		M–2

2.2 Separation and Extraction Conditions

The porphyrins of four crude oils were analysed qualitatively
and quantitatively well known methods[2]. Crude oil (0.2-
0.4g) was separated by silica gel chromatography
after removal of asphaltenes with n-pentane. The benzene fraction
was analysed qualitatively and quantitatively by visible spectrophotometry. This method can determine the content of metalloporphyrins quantitatively, but the amount of metalloporphyrins
was too low to demetal. An alternate method using dimethylformamide
(DMF) to extract the metalloporphyrins from crude oil was tried.
Crude oil samples 3-5g were extracted with 200ml aliquots of DMF
four times on a water-bath(70°C), heating for 20 minutes every time.
Afterward the extracts (DMF) were extracted with ether. The ether extract was concentrated after removal of DMF with distilled water and the
fractions were separated by silica gel chromatography. By
the two methods mentioned above, not only was the content of
metalloporphyrins determined quantitatively, but also more
porphyrins were obtained for demetallation.

Extraction and separation of source rock, oil shale and
coal:

100-250mg samples(60 mesh) were extracted with the mixed
solvent ethanol-benzene(1:9) until the extracts were colorless.
After concentration the extracts were treated with n-pentane
to remove asphaltenes and separated by column chromatography.

2.3 Demetallation with methanesulfonic acid (MSA)

The metalloporphyrins were heated with MSA (0.2-0.6ml) on a
water-bath for 1-1.5hr. at 100°C and then diluted and cooled
with ice to room temperature. The free porphyrins were extracted with ether after neutralizing to ph 7 with sodium
acetate. For purification of the free porphyrins, the ether
solution was extracted with 15% HCL and conversely the HCL

solution was extracted with ether and the procedure was
repeated 2-3 times. The ether solution containing free porphyrins was dried with Na_2SO_4 and the free porphyrins identified by visible spectrophotometry at 300-700nm.

The types of free porphyrins were identified by UV, MS
and LC and then the free porphyrins were fractionated by
LBC-1 thin-layer chromatography performed in Qingyun Factory,
Beijing. The TLC plate was coated with silica gel G it is
thickness of 2mm. Dichloromethane and hexane (1:2) were used
as eluent solvent. After the components showed clear demar-
cation lines (the colours of various ring-zones were from
deep to light red under fluorescent lamp), the polarity of
the eluent was increased a little and the components of var-
various ring-zones were collected separately. The five components were labelled A1, A2, A3, A4, and A5 according to
the sequence of elution.

2.4 Analytical Condition

Visible spectra: Specord UV/VIS spectrophotometer (W. Germany).
Fluorescence spectra: JD-30 fluorescence spectrophotometer
(France). Computerized- chromatography-mass spectrometry:
JGC-20KP/JMSD-300/SMA-2000(Japan); electron energy 70eV;
injector temp. 380°C. HPLC: SP-8100 system(U.S.A.) used for
transporting the solvents; SP-8440 system as UV detector
and silica gel(partisil, 5μm, 4.6mm x 30cm) used for
analysis. Hexane containing 1% acetic acid was solvent A,
dichloromethane solvent B and propanone solvent C. The rate
of solvent elution (A:B:C) was changed from 93:1:1 to 80:10
:10 over a 20 min. period; flow rate 2ml/min. wavelength
400nm (UV).

3. RESULTS

3.1 Visible Spectra

The absorption peaks are at 516 and 554nm in the visible
spectra of all samples, showing the absorption character of
typical nickel-porphyrins (Fig. 1). The vanadylporphyrins
were not determined. Nickel-porphyrin contents are given
in Table 2. The visible spectra of free porphyrins are
shown in Fig. 2. The demetallated porphyrins of crude oil
and oil shale have four absorption peaks in addition to the
Soret peak, that is, 550nm (IV), 534-535nm (III), 564-565nm
(II) and 616-618nm (I). The order of their intensities is
IV > II > I > III, which is typical of the character of
mixed- and deoxophylloerythroetio-porphyrins (DPEP). There
is an obvious but small peak at 598nm in the visible spectra
of crude oil, oil shale and coal porphyrins. It may be the
characteristic absorption peak of ETIO-Porphyrins[4]. The
visible spectra of five components of free porphyrins of
sample No. 5 (oil shale) are shown in Fig. 3.

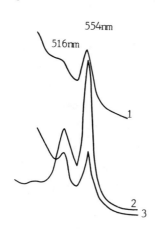

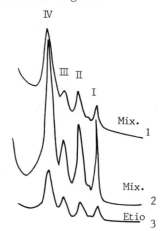

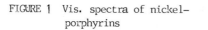

FIGURE 1   Vis. spectra of nickel-
            porphyrins

1.   Fushum soft-coal (No.9)
2.   Fushum oil shale (No.6)
3.   Jianghen crude oil (No.4)

FIGURE 2   Vis. spectra of free
            porphyrins

1.   Renqiu crude oil (No.2)
2.   Fushum oil shale (No.6)
3.   Fushum soft- coal(No.9)

TABLE 2   DATA of nickel porphyrin samples

Type No.	Area	Metalloporphyrins Content (ppm)	Carbon number of porphyrins		Methyl number porphyrins number	Most abundant	$\dfrac{DPEP}{ETIO}$	Type of vis.
1	Well Yi 20 Shengli	65.07	D	28–34	8–14	462	3.4	Mix.
			E	28–32	8–14			
2	Well 104 Renqiu	53	D	27–33	7–13	462	3.0	
			E	27–33	7–13			
3	Qianjiang Jianghei	424.4	D	27–34	7–14	462	2.2	
			E	27–34	7–14			
4	Gaosheng Liaohe	90.9	D	28–35	7–16	462	3.1/3.2[a]	
			E	28–32	8–12			
5	Fushun	14.39						DPEP
6	Fushun	17.46	D	30–35	10–15	476	4.5/4.7[a]	
			E	29–33	9–13			
7	Fushun	31.19						
8	Maoming	2.08						
9	Fushun	2.04	D	28–32	8–12	478	0.3/0.39[a]	ETIO
			E	28–36	8–16			
10	Maoming	0.25						
11	Colorado U.S.A.*	55	D	27–34	7–14	462	5.0	DPEP
			E	27–34	7–14			
12	Ajiaguli* Iran	120	D	26–38	6–18	436	0.96	Mix
			E	26–38	6–18			

*Baker and Yen (1967)   a HPLC data

Left-margin group labels: oil (Types 1–3), Crude (4), shale (5–6), Oil (7–8), Coal (9–10), shale Oil (11), Bitu (12)

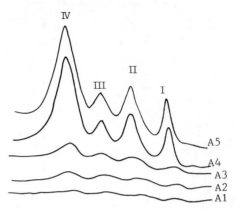

FIGURE 3 Vis. spectra of A1, A2, A3, A4, A5, components

The characteristic absorption peaks showing the main types of free porphyrins were resolved more clearly after separating by thin-layer chromatography.

The free porphyrins were separated into seven colour zones from deep-red to light-red on the silica gel coated thin-layer chromatography plate, since two of the colour zones were too small to be collected only five components were collected, that is, A1, A2, A3, A4, A5. Their vis. spectra show that they are grouped into two types of porphyrins: A1, A2, and A3 are etioporphyrins (ETIO), A4 and A5 are DPEP.

3.2. Mass Spectrometry (MS)

The mass characteristics of free porphyrins of crude oil, oil shale and coal are shown in Fig. 4-6 and Table 2.

The free porphyrins of crude oil (Fig. 4) are mainly DPEP (M^+, $m/z = 308+14n$). The ratio between DPEP and ETIO is approximately 3. The carbon numbers of free porphyrins of four crude oils are from C_{27} to C_{35} and the alkyl carbon numbers around the porphyrin nucleus are between 7-14. The strongest peak is at C_{31} (DPEP). The free porphyrins of oil

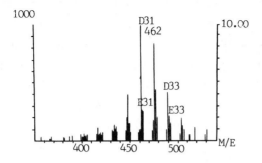

FIGURE 4   Mass spectrum of free porphyrins
of Jianghen crude oil(No.3)

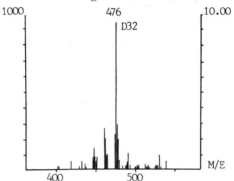

FIGURE 5   Mass spectrum of free porphyrins
of Fushum oil shale(No.6)

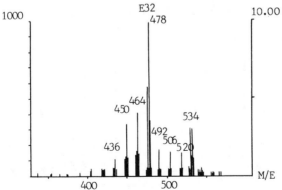

FIGURE 6   Mass spectrum of free porphyrins
of Fushum soft—coal(No.9)

shale are also mainly DPEP and their ratio of DPEP to ETIO
is 4.5, which is more than that in crude oil. The strongest
peak is C_{32}-DPEP (M^+, m/z = 476) and the carbon number range
is narrower than crude oil C_{29}-C_{35}). The character of demetallated porphyrins of coal Fig. 6 is quite different from
crude oil and oil shale. The MS shows they are mainly ETIO
and the ratio between DPEP and ETIO only 0.3. In the ETIO
series the C_{32} compound (M^+, m/z = 478) dominates.

The DPEP (A4) and ETIO (A3) mass spectra of demetallated porphyrins of sample No. 5, separated by thin-layer
chromatography, are given in Fig. 7-8. Both molecular
ranges are widened after separation by thin-layer chromatography. The change in ETIO is more obvious (Fig. 8). Its
carbon number range is from C_{24}(366) to C_{34}(506) with the
strongest peak being C_{26}. It ranges 5-6 carbons less than
C_{31}(or C_{32}) ETIO of unseparated demetallated porphyrins.
It is possible that the lighter molecules were concentrated
by thin-layer chromatography.

3.3. HPLC

Because the structures of alkylporphyrins are complicated,
it is necessary to use additional separation techniques for

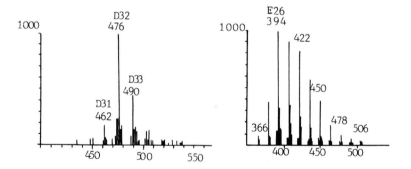

FIGURE7 Mass spectrum of A4   FIGURE 8 Mass spectrum of A5

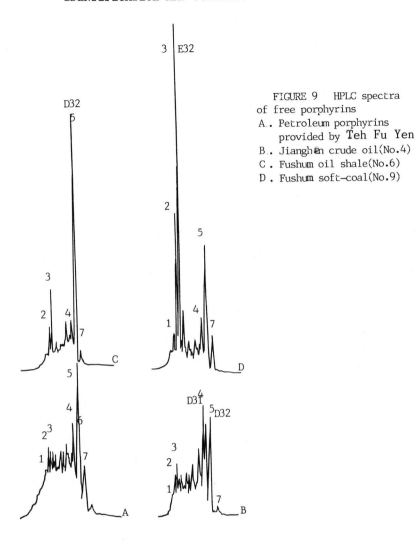

FIGURE 9   HPLC spectra
of free porphyrins
A. Petroleum porphyrins
     provided by Teh Fu Yen
B. Jianghán crude oil(No.4)
C. Fushun oil shale(No.6)
D. Fushun soft-coal(No.9)

TABLE   3   Classification of various peaks

Peaks No.	1	2	3	4	5	6	7
Classification	E_{30}	E_{30}	E_{32}	D_{31}	D_{32}	D_{30}	D_{31}

studying them. HPLC is one efficient method of separating
alkyl-porphyrins[5]. The HPLC spectra of four free base
porphyrins (sample No. 3, 6, 8 and the sample supplied by
Teh Fu Yen) are given in Fig. 9. The classification of some
major peaks is given in Table 3 according to their retention
times. Fig. 9 and Table 3 show that ETIO eluted earlier
than DPEP for all samples as is also the case in thin-layer
chromatography. The HPLC fingerprint of sample No. 3
(Jianghan crude oil) shows it consists of mixed-porphyrins
which are mainly C_{31}-DPEP. Its DPEP to ETIO ratio is 3.2.
The HPLC result for sample No. 6 (Fushun oil shale Fig. 9c)
is dominated by DPEP porphyrins with a main peak of C_{32}-DPEP.
The DPEP to ETIO ratio is 4.7. The soft coal (Fig. 9d) is
contrary to both mentioned above. Its main peak is C_{32}-ETIO
(478) and the ratio between DPEP and ETIO is only 0.39.
These HPLC results are consistent with the results of vis.
spectrometry and MS.

3.4. Fluorescence spectra

Free porphyrins emit special brilliant-red fluorescence under UV light. Thus fluorescence spectrophotometry may be
used to identify the components of free base porphyrins
after separation by thin-layer chromatography. The fluorescence spectra of five components (A1, A2, A3, A4 and A5)
of DPEP and ETIO are all at Em 573 nm and Em 625 nm (Fig. 10)

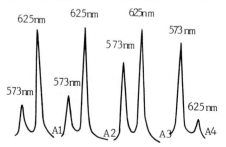

FIGURE 10 Fluorescence
emitting spectra of A1,
A2, A3, A4

similar to the vis. spectra of DPEP and ETIO, the position
of the peaks was unchanged, but the intensity of each peak
changed with type. The ratio of the intensity of two fluorescence emission peaks changed regularly according to the sequence of elution by thin-layer chromatography (Table.4).
The ratio between 573 and 625 nm of ETIO is smaller than
that of DPEP, but it is necessary to carry out studies before the ratio can be used to identify either DPEP or ETIO.

4. Discussion

    1. The nickel porphyrins and not vanadyl porphyrins in
crude oil, oil shale and coal are analyzed in this study.
Few vanadyl porphyrins were found in the continental sediments of China until present. Vanadyl porphyrins were found
in a sample collected from the Jurassic formation in

TABLE 4   Date for various components of free porphyrins

Com.	Vis. spectra nm	FSE* nm	Em573 Em625	mass spectra	Note
A1	500, 531.9, 568.7, 618	400		366,380,394,408,	ETIO
A2	500,531.9,568.7,619	400	0.21	422,436,450,464,	310+14n
A3	500,532.6,566.9,617	400	1.45	478,492,506	
A4	504,535,567,619	400	2.57	434,448,462,476,	DPEP
A5	505,540,571,620	400	5.59	490,504,518,532	308+14n

FSE* : Fluorescence Spectra Ex.

the Aicen 1 well, Xinjiang, but their occurrence is too
small for quantitative determination (Fig. 11). This is a
characteristic of porphyrins in the continental sediments
of China. Until present, the sample with the highest petroleum porphyrins content in the continental sediments of
China, is Fushum oil shale from the tertiary (Eocene). Its
content of nickel porphyrins is 31.1 ppm (in the rock). It contains sev-

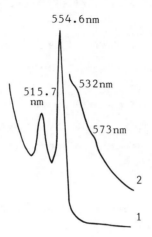

FIGURE 11 Vis. spectra of nickel and vanadyl porphyrins of Aicun 1
well ,Xinjing(2459 m, deep-grey mudstone)
1.Ni Porphyrin   2.V=O Porphyrin

eral hundred times higher porphyrin content of the source
rocks of Qianjiang Formation (Tertiary) of Jianghen basin
(0.052 ppm), and nearly 10 times as much as that in the oil
shale in Colorado, U.S.A. (Table 2).   Thus, it is obvious
that the continental sediments of China are characterized by
a low porphyrin content.

2.   In all samples analyzed, the alkyl-porphyrin and
DPEP, ETIO, RHODO and Di-DPEP, but the abundance ratios and
structures of them are different in crude oil, oil shale and
coal..

In general, the alkyl-porphyrins in Chinese continental
crude oils show DPEP as the dominant type and ETIO as the
second.   However, it has been suggested that DPEP could
change to ETIO with increase of maturity of crude oil and
sedimentary rocks.   The free porphyrins of crude oil analyzed
in this study are of mixed type according to the vis. spectra and the ratio between DPEP and ETIO is 2.2-3.4.   The
free porphyrins of fushun oil shale are of the DPEP type and

the ratio between DPEP and ETIO is 4.5. The free porphyrins of coal belong mostly to the ETIO series, the DPEP to
ETIO ratio being only 0.3.

The most abundant free porphyrin in crude oil analyzed
in this study is C_{31}-DPEP (molecule weight 462); Fushum oil
shale C_{32}-DPEP (476) and coal C_{32}-ETIO (478).

The carbon number range of porphyrins in crude oil is
between C_{27} and C_{36}, and the alkyl carbon atoms outside the
porphyrin nucleus total 7-14[8]; oil shale C_{24}-C_{25} and its
alkyl carbons 9-15; coal C_{28}-C_{36} and its alkyl carbons 8-16.

From the three points mentioned above, it is seen that
the type and structure of porphyrins are quite different
among crude oil, oil shale and coal. The free porphyrins of
soft-coal from Fushum are mainly ETIO. Palmer et al[6] thought
it possible that the ETIO series of porphyrins in coal found
in the early stage of coalification was formed due to degen-
eration of the DPEP series which still remains in the late
stage. However, the highest carbon number of this porphyrin
is C_{36} and this is different from America's coal (Palmer et
al[6]) which has a carbon number of not over C_{32}. The kerogen
of Fushum coal is 80-90% vitrinite determined microscopically
and it is deep-brown. Its IR spectrum shows that there
are a few aliphatic structures and more aromatic structures.
The aromatic polymerization is strong and the absorption
bands of functional groups containing oxygen are also strong.
The H/C and O/C (atomic ratios) of this sample are 0.93 and
0.227 respectively and its oxygen content is the highest
among all oil shales and coals analyzed (Table 5). These
results show that its ETIO porphyrins formed in the early
stage of coalification are not totally derived from DPEP,
because this sample (F-4) is low mature and the type of
porphyrins is mainly ETIO. This view was expressed recently

by Barwise and Roberts[7] with respect to porphyrins in oil shale.

TABLE 5.   Elemental data, IR of kerogen of oil shale, coal

Samples	C%	H%	N%	S%*	O%	H/C	O/C	$\dfrac{2900cm^-}{1600cm^-}$	$\dfrac{1700cm^-}{1600cm^-}$
Fushun F-2	75.65	9.04	2.40	1.35	11.56	1.43	0.155	2.07	1.10
Fushun F-3	71.82	7.94	2.73	1.56	15.95	1.33	0.167	1.55	0.88
Fushun F-5	69.46	8.40	2.35	0.96	18.81	1.45	0.203	1.84	0.96
Máoming M-1	73.74	8.22	2.69	2.05	13.31	1.34	0.135	1.64	0.96
Maoming M-2	70.54	5.41	2.83	1.01	20.21	0.92	0.215	0.64	0.81
Fushun F-4	69.62	5.39	3.43	0.53	21.03	0.93	0.277	0.61	0.79

* Organic sulfur

3. The free porphyrins of sample No.5 were separated into five components by thin-layer chromatography. The vis. spectra of A1, A2, and A3 indicate ETIO porphyrins having a wide number range (C_{24}-C_{34}). In this ETIO series, the most abundant porphyrin is C_{26}-ETIO (m/z 394). It is possible that the ETIO having a relatively small molecular weight were concentrated because the ETIO porphyrins were enriched at the fore edge of silica gel plate due to separation by thin layer chromatography. Thin-layer chromatography is effective in the purification of free porphyrins and it provides a useful means for the study of porphyrin type and structure.

ACKNOWLEDGEMENT
Our sincere thanks are due to coworkers Li Yimin and Zhang Ling and to Zhang Dajiang for providing the investigated samples.

REFERENCES

1. YANG ZHIQIONG, (Report on petroleum Exploration
    Research, Beijing,1964) vol.3 , 46–53

2. HSU LINE, Geochimica (Chain),3,174–185,(1973)

3. SHI JIYANG, Scientia Sinica(series B)China, 11,1019–1026,(1982)

4. E. W. BAKER, T. F. YEN, Journal of American Chemical Society,
    89, 3631(1967)

5. G. Eglinton, S.K. HAJIBRAHIM, J.R. MAXWELL, and J.M.E. QUIRKE
    Advances in Organic Geochemistry (1979) ,A.G.DOUGLAS AND J.R.
    MAXWELL,EDS., Pergamon press,193–203(1980)

6. S.E.PALMER, E.W.BAKER, L.S.CHARNEY AND L. W. LOUDA Geochimica
    et Cosmochimica Acta 46, 1233–1241(1982)

7. A.J.G. BARWISE AND I.ROBERTS, Organic Geochemistry 6, 167–176
    (1984)

8. E.W.BAKER, S.E.PALMER, The Porphyrins Academic press,Inc.
    New York vol.1 500–501(1978)

STUDIES ON THE STRUCTURE OF THE TERPENOID SULFIDE TYPE BIOLOGICAL MARKERS IN PETROLEUM

JOHN D. PAYZANT, TERRY D. CYR, DOUGLAS S. MONTGOMERY AND OTTO P. STRAUSZ*

Department of Chemistry, University of Alberta, Edmonton, Alberta, Canada T6G 2G2.

The structure of the major C_{13} component of the previously described bicyclic terpenoid sulfides, probably the first sulfur containing biological marker molecules detected in petroleum has been identified as:

Similarly, the most probable structure for the major C_{23} component of the tetracyclic terpenoid sulfide series is:

A general procedure for the isolation of sulfides from petroleum is described. The key steps are the selective oxidation of sulfides to sulfoxides with photochemically generated singlet oxygen and the subsequent regeneration of the sulfides by reduction with lithium aluminium hydride.

---

*To whom correspondence should be addressed.

1. INTRODUCTION

In the past, considerable interest has been shown in the chemistry of the sulfur compounds present in petroleum[1-3]. The sulfur-containing molecules previously reported in petroleums and sediments have been usually low molecular weight or aromatic compounds resulting from maturation processes[4]. Recently, we reported the discovery of a bicyclic and a tetracyclic homologous series of terpenoid sulfides and sulfoxides in the heavy oils of Northern Alberta[5] and, subsequently, other tri- and hexacyclic terpenoid sulfides[6-8]. Altogether these sulfides comprise a few hundred novel individual molecules, none of which had been previously reported in the chemical literature. In addition, in 1984, a thiophene ring-containing C_{35} hopanoid was identified in sediments[9].

The most abundant of the terpenoid sulfides were one of the C_{13} and one of the C_{23} isomers in the bicyclic and tetracyclic series, respectively, for which structures 1 and 2 were considered to be the most likely.

The reasons for these structural assignments were: 1) high resolution mass measurements of the molecular ions gave the elemental formulae $C_{13}H_{24}S$ and $C_{23}H_{40}S$ for 1 and 2, respectively; 2) reduction of the sulfides with Raney nickel gave the hydrocarbons 1a and 2a as was shown by capillary gc

retention time and mass spectral comparison[10-14]; 3) the
mass spectra of 1 and 2 and all members of their respective
homologous series featured a medium intensity molecular ion
along with base peaks corresponding to m/z = 183 and 319
from the dealkylated ring structures. These observations
establish that the sulfur is attached at the C_2 position of
the side chain of the hydrocarbons 1a and 2a. However, the
site of attachment of the residual valence of the sulfur remained ambiguous. We favoured the participation of the sulfur in a five membered ring, giving structures 1 and 2, over
other possible six membered ring structures. The mass spectra of 1 and 2 contain the m/z = 123 ion and 2 also has a
significant m/z = 191 ion. The former fragment ion has been
considered to be diagnostic for the presence of the trimeth-
ylated A ring of bi- and higher cyclic terpanes and the latter, of the tetramethylated AB rings of tri- and higher cyclic terpanes. In order to establish the structure of the
$C_{13}H_{24}S$ compound unequivocally we followed two parallel
routes: (1) the synthesis of two stereoisomers of 1, and (2)
the isolation of the C_{13} sulfide from the petroleum for
spectroscopic studies.

2. EXPERIMENTAL

2.1 Synthesis of the C_{13} terpenoid sulfide 1

β-Cyclocitral 3 was prepared from commercial citral as previously described[15]. Sodium ethoxide[16] catalyzes the slow
condensation and subsequent cyclization reaction between 2-
mercaptoethylacetate 4 and β-cyclocitral to afford the alcohol 5 as outlined in Scheme I. The condensation reaction
was conducted with neat reagents and is slower in protic
solvents such as ethyl alcohol. The use of excess base or

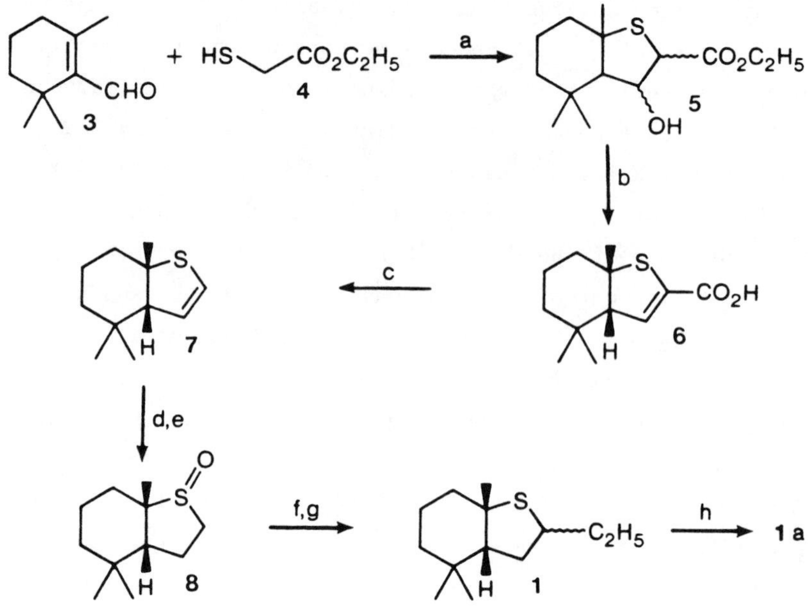

SCHEME 1. a) NaOEt (dry, 0.2 eq.) (4/3) = 1.2, 25°, Ar, 13
          days (80%); b) KOt-Bu/HOt-Bu, 25°, 16 h (48%); c)
          235°, 1/2 h (91%); d) PtO$_2$, EtOH, H$_2$ (1 atm), 5
          days (65%); e) NaIO$_4$, EtOH, H$_2$O (90%); f) n-BuLi,
          THF, -78°, 1 h, EtBr, -78° to 25°, 1 h (80%); g)
          LiAlH$_4$, dioxane, reflux 1 h; h) Raney nickel,
          EtOH, reflux 3 h.

the application of heat results in the formation of complex
mixtures. Treatment of crude 5 with two equivalents of
KOt-Bu in HOt-Bu resulted in the formation of the acid 6 in
48% isolated yield on the 0.5 mole scale. 6 slowly crystal-
lizes from a pentane solution of the crude acid product.
Considerably improved yields of 6 were obtained on a small
scale employing chromatographic purification of starting
materials and products.

The acid 6 appears to be appropriately functionalized
for subsequent conversion to 1. However, we were unable to con-
vert the carboxyl group of 6 to an alkyl group, presumably

because of the β position of the carbonyl group to the sulfur atom[17]. Typically, a reaction sequence involving this site produced material resulting from rearrangement or polymerization. However, at 235°C, 6 smoothly decarboxylates to give the olefin 7 in high yield. Catalytic hydrogenation of the double bond of 7 and subsequent oxidization gave the sulfoxide 8 which could be readily alkylated α to the sulfur atom and reduced to a mixture of two stereoisomers of 1. The spectral properties of the key synthetic intermediates have been reported[6].

Reduction of 1 with Raney nickel gave the hydrocarbon 1a. The acidic proton of the sulfoxides of 1 was exchanged for deuterium (NaH in CD_3SOCD_3, 25°, 3 h) and the sulfoxides reduced ($LiAlD_4$, dioxane, reflux 1 h) to the sulfides. The mass spectra of these sulfides, Figure 1, show the incorporation of one atom of deuterium in each isomer. As may be seen from Figure 1, the mass spectra of 1 and that of the natural material are quite different, as are the gc retention times. Therefore it is evident that the structure of the natural material is different from 1.

2.2 Isolation of the C_{13} sulfide from petroleum

In this connection we first describe a general procedure for the isolation of sulfides from petroleums. To this end a number of methods have been employed[4,18]; however, most are unsatisfactory in that either they produce samples which are impure, or the method discriminates between sulfur in different sulfide environments, or the method fails to discriminate between sulfide sulfur and thiophenic sulfur. The present procedure overcomes these problems by oxidizing the sulfides to the highly polar sulfoxides with photochemically generated singlet oxygen[19] followed by the removal of the polar

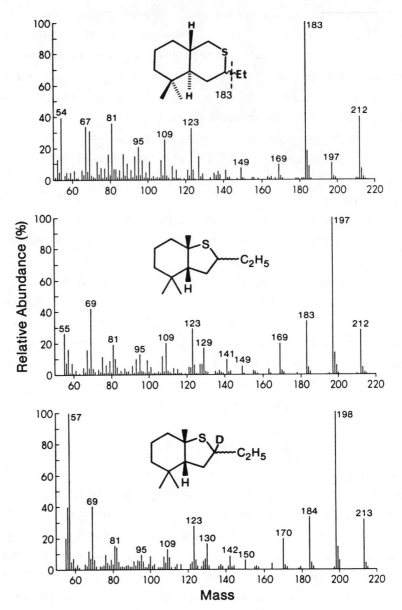

<u>FIGURE 1</u>. Mass spectra of the C_{13} terpenoid sulfide of pet-
roleum (upper panel), synthetic C_{13} sulfide $\underset{\sim}{1}$ (middle panel)
and deuterated synthetic C_{13} sulfide (lower panel) showing
one hydrogen on the carbons adjacent to the sulfur.

materials from the mixture by filtration on silica gel. Reduction of this polar fraction with $LiAlH_4$ converts the sulfoxides back to low polarity sulfides which are readily separated from the mixture.   The oxidation with singlet oxygen
does not discriminate between sulfur in different sulfide
environments, is semi-quantitative, does not oxidize sulfoxides to sulfones, and does not oxidize the sulfur in dibenzo-
thiophene as was shown in control experiments.

The sulfide isolation procedure is illustrated in the
following discussion using deasphalted heavy oil from the
Athabasca deposit of Northern Alberta.   Athabasca maltene
(10 g) was added to a 1 L 3-neck flask equipped with magnetic
stirring bar, an efficient reflux condenser and an oxygen
bubbler, together with methylene blue (0.2 g), chloroform
(400 mL), and methanol (200 mL).   Oxygen, dried by passing
it through a column of 3Å molecular sieves, was introduced
via a Pasteur pipette under the surface of the liquid at a
rate of 20 mL min^{-1}.   A 200 W incandescent bulb was suspended
a few centimetres from the flask and the whole surrounded
with aluminium foil.   The reaction mixture refluxed gently
under these conditions.

After four to five days, the reaction mixture was concentrated and applied to 150 g of silica gel.   Elution with
CH_2Cl_2 (250 mL) and 10% $EtOAc/CH_2Cl_2$ (500 mL) eluted material
of low polarity, while subsequent elution with 10% $CH_3OH/$
CH_2Cl_2 (500 mL) yielded the polar fraction containing the
sulfoxides. This fraction is a sticky, viscous liquid when
"fresh", but becomes a brittle solid on standing. This polar
fraction was reduced with excess $LiAlH_4$ in refluxing dioxane
for one hour. After a conventional workup with aqueous acid,
the organic material was filtered through 50 g of silica gel
with 25% toluene/hexanes (300 mL) to yield the crude sulfides

0.53 g (5.2% of Athabasca maltene). These sulfides are usually dark due to the presence of a small amount of polymeric material. This dark material is easily removed by a second filtration on silica gel to give the sulfides as a yellow to orange oil which darkens on standing.

The oxidation is about 85% complete under the conditions described above. The use of smaller quantities of methylene blue or the omission of methanol as cosolvent greatly slows the oxidation. This procedure has been used on a scale from 1 g to 1.5 kg and samples with sulfide contents from 0.3% to 16%.

Employing the above described procedure we used 7 kg of oil (Lloydminster, Alberta) for the isolation of sufficient quantity of the C_{13} terpenoid sulfide for NMR analysis. The oil was distilled and the lower boiling 20% was oxidized with photochemically generated singlet oxygen after which the sulfoxides were removed by filtration on silica gel and reduced back to the sulfides as described above. The sulfides were fractionally distilled through a 50 cm Vigreaux column and this procedure was monitored by gc. The appropriate sulfide fractions were chromatographed on silica gel eluting with 15% toluene/hexane. This afforded samples in which the concentration of the C_{13} terpenoid sulfide was a few percent. Further chromatography on a variety of supports failed to significantly enhance the purity; however, reverse phase chromatography on C_8 bonded silica gel (EtOH/H_2O, 95:5) yielded a sample of 55-60% purity.

2.3. ^{1}H NMR and deuterium exchange experiments
The 400 MHz ^{1}H NMR spectrum of the C_{13} sulfide sample isolated from petroleum, Figure 2, reveals the presence of only three methyl groups in the molecule, two singlets and a

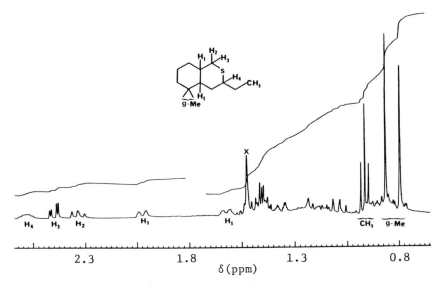

FIGURE 2.   400 MHz ^{1}H NMR (CDCl$_3$) of the C$_{13}$ terpenoid sul-
            fide from Lloydminster heavy oil. X indicates
            impurity. The integration curves suggest that
            the sample is 55–60% pure.

triplet, along with three one–hydrogen–atom signals assignable to hydrogens on the carbons adjacent to the sulfur and
two one–hydrogen–atom signals assignable to the ring junction
hydrogens of 9.

The remaining signals appear as a complex pattern. This
sulfide sample was oxidized (HOAc, 30% H$_2$O$_2$, 25°, 16 h) to
the corresponding sulfone. In the ^{1}H NMR of the sulfone,
Figure 3, the signals assigned to the hydrogen atoms on the
carbon atoms adjacent to the sulfur shift significantly,
while the remaining signals are much as before.

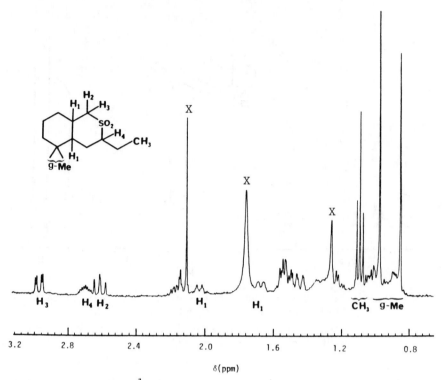

FIGURE 3.   400 MHz ^{1}H NMR (CDCl$_3$) of C$_{13}$ terpenoid sulfone
from Lloydminster heavy oil. X indicates
impurities from oxidizing reagents.

The acidic protons of the sulfone may be exchanged for
deuterium ($\underline{n}$–BuLi/CD$_3$SOCD$_3$, 25°, 16 h, D$_2$O quench). The mass
spectra of the sulfone and its deuterated analogue are shown
in Figure 4 where three atoms of deuterium have been incorporated.

Chromatography of the crude sulfide fraction from Athabasca bitumen on silica gel (15% toluene/hexanes) easily af-
fords a sample of the C$_{23}$ tetracyclic terpenoid sulfide of
sufficient purity for good GC-MS analysis. The C$_{23}$ sulfide
was oxidized to the sulfone and the acidic hydrogen atoms

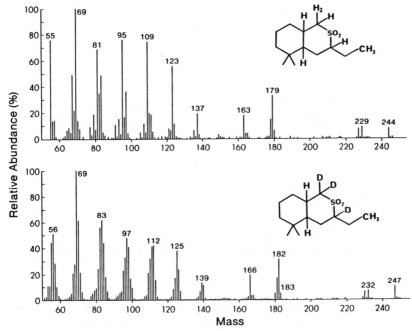

FIGURE 4.   Mass spectra of the C_{13} terpenoid sulfone from
            Lloydminster heavy oil (upper panel) and the cor-
            responding deuterated sulfone (lower panel) show-
            ing the incorporation of three atoms of deuterium.

exchanged for deuterium using the same conditions as described for the C_{13} sulfide. The mass spectra of the C_{23}
sulfone and its deuterated counterpart are shown in Figure 5
and indicate the incorporation of three atoms of deuterium.

3. DISCUSSION

The mass spectrum of the C_{13} terpenoid sulfide (upper panel,
Figure 1) with the base peak at M-29 is consistent with the
sulfur atom being attached to the second carbon atom of the
n-butyl side chain of 1a. The deuterium exchange experiments
and NMR data show three hydrogens on the carbons adjacent to
the sulfur and the NMR spectra reveal the presence of three

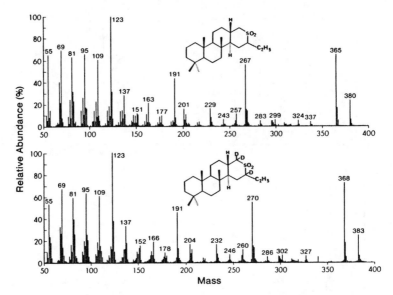

FIGURE 5. Mass spectra of C_{23} tetracyclic terpenoid sulfone
from Athabasca heavy oil (upper panel) and
corresponding deuterated sulfone (lower panel)
showing the incorporation of three atoms of
deuterium.

methyl groups, the multiplicities of which are satisfied by
9.

The ms fragmentation pattern of the two different ring
systems 1 and 9 show characteristic differences in Figure 1.
When the sulfur resides in a six membered ring, the $(M-R)^+$
at $m/z = 183$ is the base peak for the entire series, while
in the case of the five membered ring it is the $(M-15)^+$ ion.
Therefore we postulate that the entire homologous series of
bicyclic terpenoid sulfides we reported to occur in petroleum has the ring framework of 9 and, by inference, the main
homologous series of tetracyclic terpenoid sulfide should
have an analogous ring structure 10.

The ms fragmentation patterns for the two series are similar, $(M-R)^+$ at $m/z = 319$ being the characteristic base peak for the entire tetracyclic series. The C_{23} tetracyclic sulfone incorporated three atoms of deuterium under the same conditions used for the C_{13} sulfone. The mass spectra of the sulfones, Figure 5, give the characteristic $m/z = 191$ ion for the deuterated and undeuterated compound consistent with structure 10.

These series of bicyclic and tetracyclic terpenoid sulfides are widespread in the petroleums of the Western Hemisphere[8]. In most samples, the C_{13} compound 9 is the most abundant member of the bicyclic series and the C_{23} compound 10 is the most abundant member of the tetracyclic series. The bicyclic series extends from C_{11} to C_{27} while the tetracyclic series extends from C_{21} to C_{40} in some samples. Recently, we[20] reported the occurrence of a series of hopane sulfides in which the sulfur atom is also attached to the second carbon atom of the side chain forming a five or a six membered ring. Thus, all three main series of terpenoid sulfides found in petroleum to date have this structural feature in common, which may reflect the site specificity of the biosynthetic pathway responsible for the sulfur incorporation.

4. ACKNOWLEDGEMENTS
The financial support of the Alberta Oil Sands Technology

and Research Authority and the Natural Sciences and
Engineering Research Council of Canada is gratefully
acknowledged.

REFERENCES

1. W.L. ORR, Biogeochemistry of Sulfur, in Handbook of
   Geochemistry, Edited by K.H. Wedpohl, Vol.II-I,
   Sec. 16-L (Springer-Verlag, Berlin, 1974).
2. W.L. ORR, Sulfur in Heavy Oils, Oil Sands and Oil
   Shales, in Oil Sand and Oil Shale Chemistry, Edited by
   O.P. Strausz and E.M. Lown (Verlag Chemie, New York,
   1978) pp.223-243.
3. B.P. TISSOT AND D.H. WELTE, Petroleum Formation and
   Occurrence (Springer-Verlag, 1984) pp.398-401, 411-414.
4. H.T. RALL, C.J. THOMPSON, H.J. COLEMAN AND R.L. HOPKINS
   Sulfur Compounds in Crude Oil, U.S. Bur. Mines, Bull.
   659, pp.1-187 (1972).
5. J.D. PAYZANT, D.S. MONTGOMERY AND O.P. STRAUSZ, Tetrahedron Lett., 24, 651-654 (1983).
6. J.D. PAYZANT, T.D. CYR, D.S. MONTGOMERY AND
   O.P. STRAUSZ, Tetrahedron Lett., 26, 4175-4178 (1985).
7. J.D. PAYZANT, A.M. HOGG, D.S. MONTGOMERY AND
   O.P. STRAUSZ, AOSTRA J. Res., 1, 203-210 (1985).
8. J.D. PAYZANT, D.S. MONTGOMERY AND O.P. STRAUSZ, Org.
   Geochem., 1986, in press.
9. J. VALISOLALAO, N. PERAKIS, B. CHAPPE AND P. ALBRECHT,
   Tetrahedron Lett., 25, 1183-1186 (1984).
10. C.M. EKWEOZOR AND O.P. STRAUSZ, Tetrahedron Lett.,
    23, 2711-2714 (1982).
11. T.D. CYR AND O.P. STRAUSZ, J. Chem. Soc., Chem. Commun.
    1028-1030 (1983).
12. F.R. AQUINO NETO, A. RESTLE, J. CONNAN, P. ALBRECHT AND
    G. OURISSON, Tetrahedron Lett., 23, 2027-2030 (1982).
13. D. HEISSLER, R. OCAMPO, P. ALBRECHT, J.J. RIEHL AND
    G. OURISSON, J. Chem. Soc., Chem. Commun., 496-498,
    (1984).
14. M.G. SIERRA, R.M. CRAVERO, M.A. LABORDE AND E.A. RUVEDA,
    J. Chem. Soc., Chem. Commun., 417-418 (1984).
15. L. COLOMBO, A. BOSSHARD, H. SCHINZ AND C.F. SEIDEL,
    Helvetica Chemica Acta, 34, 265-273 (1951).
16. C.D. HURD AND L.L. GERSHBEIN, J. Am. Chem. Soc., 69,
    2328-2335 (1947).
17. C.R. NOLLER, Chemistry of Organic Compounds (Saunders,
    London, 1966), p.316.

18. G.D. GAL'PERN, Russian Chemical Reviews, 45, 701-720
    (1976).
19. C. GU, C.S. FOOTE AND M.L. KACHER, J. Am. Chem. Soc.,
    103, 5949-5951 (1981).
20. T.D. CYR, J.D. PAYZANT, D.S. MONTGOMERY AND
    O.P. STRAUSZ, Org. Geochem., 9, 1939-143 (1985).

PORPHYRIN TYPES AND EPIMERS OF TRITERPANES AND
STERANES IN COALS OF DIFFERENT RANKS IN SOUTH-
EASTERN UINTA REGION

YONG-DA GU*, YU WANG** and TEH FU YEN
University of Southern California
Los Angeles, CA 90089-0231

Two samples of coal from Southeastern Uinta region,
only 20 miles apart, yet appearing to be of
different ranks, were characterized with biomarker
studies. From porphyrin types, metal distribution and
other related biomarker parameters, the Crested Butte
coal is typically more mature and the Grand Mesa coal
is less mature. The presence of cadalene, retene and
simonellite and the absence of acenaphthenes and
fluorenes in Grand Mesa coal may indicate a mild
catagenesis. Triterpane and sterane studies show
that during diagenesis the Grand Mesa sample was
exposed to an anarobic, less watery environment, whereas the Crested Butte sample was exposed to an aerobic,
aqueous environment with more bacterial reworking. It
appears that Crested Butte coal encountered some heat
and pressure during its formation.

* Present address: Institute of Coal Chemistry,
  Chinese Academy of Sciences, Taiynan, Shensi, PRC.
** Present address: Department of Environmental
  Engineering, East China University of Chemical
  Technology, Shanghai, PRC.

1. INTRODUCTION

The coal fields of the Uinta region in Colorado occupy the
moderately to steeply dipping edge of the Piceance Creek
basins. Although apparently a simple regional structure,
it is one which is heavily modified by faults, folds
and intrusions covering various local areas of structural
complexity. Physiographically, the Uinta region is located
in the Rocky Mountain region neighboring a number of coal
producing areas where most coal resources of the Western
U.S.A. originate (Fig. 1).

A small region on the South-East edge of Uinta basin
is of particular interest. The Crested Butte quadrangle
(Fig. 2) lies in a site of merging provinces by the Elk
Mountain range separated from the plateau province. This
region is surrounded by large igneous masses and has some
evidence of laccolithic intrusion.

What is more unusual is that coals of different ranks
are found including subbituminous, high volatile C
bituminous, semi-anthracite, and even meta-anthracite.
Two coal samples are selected for study, 26L is a high
volatile bituminous coal from Crested Butte whereas 51I is
a subbituminous coal from Grand Mesa field, about 15 miles
southwest from Crested Butte field (See Fig. 2 and 1).
Both samples differ greatly in their chemical and
physical properties (Table 1).

According to Landis (1), the coals belong to the
Messaverde Formation of Late Cretaceous age. Difference in
ranks may be due to metamorphic changes modified through
geological events such as temperature and pressure in
subsequent time (2). Climatic differences are in the
extremes even today, Grand Junction (altitude 4586 ft), the

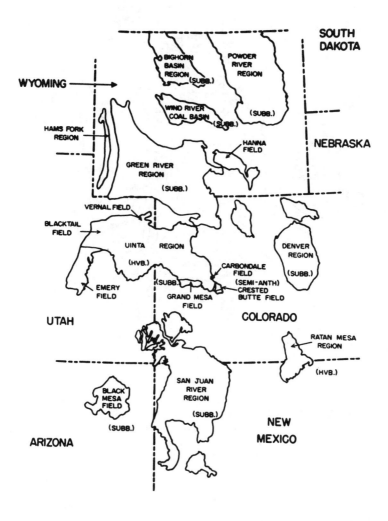

FIGURE 1   Coal fields near the Uinta Region; dotted lines
are state boundaries; rank of coal is indicated
in parentheses under various coal fields.

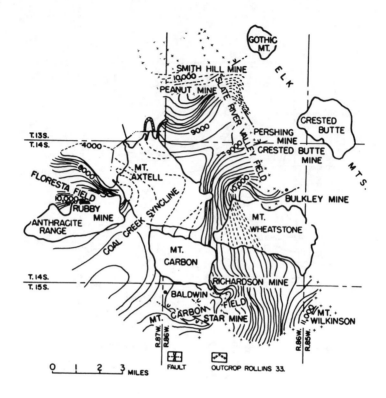

FIGURE 2   Map of Crested Butte Quadrangle; Grand Mesa is
             Ca. 15–20 miles south east from it.   Contour
             interval 100 ft. The structure of the Rollins
             sandstone is based on Ref. 2.

largest city in western Colorado, has an average annual
precipitation of 8.86 inches; whereas Crested Butte
(altitude 8885 ft), has an average annual precipitation of
27.28 inches. The constant weathering and errosion in the
Crested Butte region is an important factor.

    Although those coal deposits occur close to each

other, their rank, type and properties are distinctly
different. Their biomarkers were studied in order to
understand further the geochemical significance of their
formation.

2. EXPERIMENTAL

Coals upon receipt are kept in Mylar bags and stored
in the refrigerator. For convenience we shall refer to the
5lI coal as Grand Mesa and the 26L coal as Crested Butte.
The Grand Mesa is shiny in appearance whereas the Crested
Butte is quite dull. Their properties are listed in Table 1.
Both coals were ground and sized to 60 mesh before
Soxhlet extraction in 100g quantities using a benzene-
methanol azeotropic mixture. The extract solvent was
removed to recover the bitumen. The bitumen was
subsequently eluted with n-hexane, toluene, and toluene-
methanol (1:1) successively from a silica gel column.
The first fraction (n-hexane) was used for GC and GC-MS
studies for n-alkanes, isoprenoids, triterpanes, steranes
and polynuclear aromatic hydrocarbons. The second fraction
was used for porphyrin and metal studies.
A Hewlett Packard model 5880A GC-FID with an OV-101
fused silica capillary column was used in this work. The
injection temperature were 250°C with He as carrier gas, and
a flow rate of 3 ml/min. Two types of temperature program
were used: 50 - 120°C at 10°C/min rate (held for
2 min. at 50°C) and 120 - 250°C at 6°C/min (held for 5 min
at 120°C).
For GC-MS, Finnegan 4000 system with Incos 2300 data
system was used. A DB5 capillary column of 30 m x 0.25
mm ID with a temperature program of 35-280°C at 4°C/min

and 280–310°C at 2°C/min was used.

TABLE I   Chemical and Physical Properties of South
            Eastern Uinta Region Coals.

No.	51I	26L
Location	Grand Mesa	Crested Butte
Rank	Subbituminous A	High Volatile Bituminous B
Type	Vitrinite 82.3%	Dull coal 11.1%
Age	Cretaceous	Cretaceous
Elemental Analysis (DMMF):		
%C	73.84	84.86
%H	4.58	5.36
%N	0.77	1.66
%S	0.20	0.65
%O	20.60	7.87
H/C	0.75	0.76
O/C	0.21	0.07

High resolution mass spectrometry was done on an AEI
MS 902 with an electron beam energy of 58 eV using direct
solid insertion probe at 250°C.

For HPLC separations a Waters Associates model
6000A solvent delivery system in conjunction with model
440 absorbance detector set at 405 nm. was used. A normal phase
silica gel column with a solvent gradient of 2 - 50%
chloroform in hexane as the mobile phase was used at a
flow rate of 1.8 ml/min.

Dematallation of porphyrin was accomplished by the
usual methanesulfonic acid (3).

3. RESULTS AND DISCUSSION

Porphyrin type analysis based on mass spectra data (Fig. 3) indicates that Crested Butte yields a simpler spectrum than that of Grand Mesa. For Crested Butte (Fig. 3a) only Etio type (310 + 14n) homologs are present, with no detectable DPEP type (308 + 14n) homologs. Furthermore, the homologous series is short (only $C_{28} - C_{30}$) and truncates on the low molecular weight side with predominance of porphyrin of C_{28} group (m/z 422). All Etio series show a high degree of dealkylation.

In contrast, the Grand Mesa (Fig. 3b) contains both Etio and DPEP series of porphyrins and the spread is quite broad. The highest intensity at m/z 478 is an etioporphyrin. The DPEP series is weak but is discernible (e.g., m/z 406, 420, 448, 462, 490, etc.). The DPEP/Etio envelop ratio (4) computed for Grand Mesa is 0.1.

Consistant with the mass spectral interpretation HPLC studies (Fig. 4) also show a predominant Etio series in Crested Butte, although after magnification of the Etio region (Fig. 4, peaks 1,2 and 3), peaks due to DPEP porphyrins (Fig. 4, peaks 5 & 6) are detectable. Under these HPLC conditions, Peak 1 has proven to be etioporphyrin III by coinjection, similar to published findings (5,6). Peaks 5 & 6 are identified as DPEP by coinjection with the C_{32} DPEP isolated from Boscan petroporphyrins. An approximation of the DPEP/Etio ratio for Crested Butte yielded 0.03. However, a similar approximation for Grand Mesa based on this method is approximately 0.2. The lower DPEP/Etio value indicates high thermal stress, since DPEP shifts to Etio for both increasing depth and age of the samples (7,8).

UV-visible spectra of dematallated porphyrins from

Grand Mesa are similar to those from Crested Butte (Fig. 5).
However, atomic absorption using microprobe analysis of the
metalloporphyrins indicates that vanadium is the major

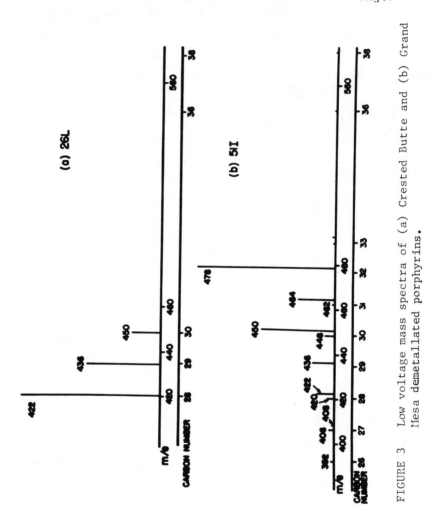

FIGURE 3   Low voltage mass spectra of (a) Crested Butte and (b) Grand
Mesa demetallated porphyrins.

constituent of Grand Mesa, whereas the Crested Butte
contains gallium and nickel (Table II).   The presence of
gallium in coal porphyrin is not rare, the existence of

TABLE II   Metals in Matalloporphyrin and Metal Free Porphyrin (ppm).

Methods	Grand Mesa		Crested Butte	
	Metalloporphyrin	Metal-Free Porphyrin	Metalloporphyrin	Metal-Free Porphyrin
	AA *	AA	MP**	AA
Cr	0.44	0.009	0.02	0.013
Co	0.092	<0.001	-	0.001
Cu	0.064	0.11	-	0.054
Ni	0.043	0.048	0.21	0.077
V	0.98	<0.012	0.02	<0.012
Ga	-	-	0.20	-

* Atomic absorption analysis.
** Microprobe analysis for EDS.

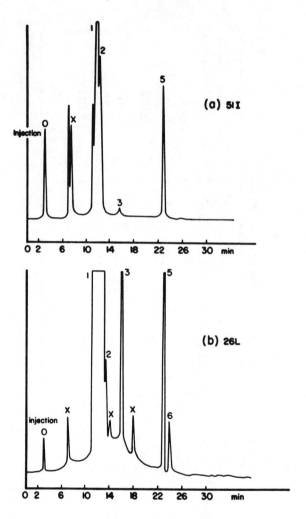

FIGURE 4   HPLC of the metal-free porphyrins from (a)
             Grand Mesa and (b) Crested Butte. Peaks 1, 2
             and 3 are Etio type and peaks 5 and 6 are
             phyllo type. Peak 1 is identified as
             Etioporphyrin III by coinjection,

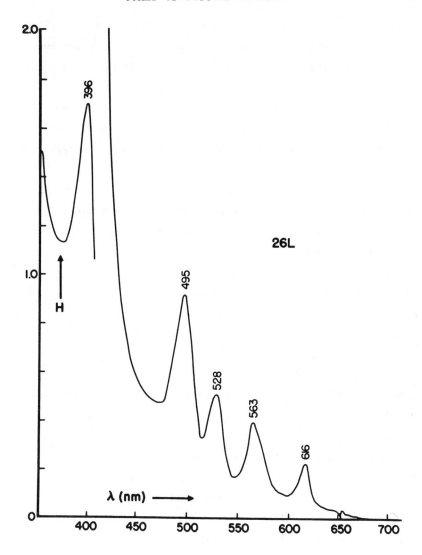

FIGURE 5   Visible spectrum of Crested Butte demetallated
            porphyrins.

gallium porphyrins for a few British bituminous coals and
one Turkish lignite is known (9). Yet, to date, the

geobiological function of gallium is not understood. It
appears that under a set of wet, microbiological aerobic
conditions, porphyrin can chelate to metals other than the
vanadyl group in the contacting brine.

Applying the commonly adopted GC and computerized GC-
MS techniques to the **n-hexane fractions of the two coal**
samples, general distributions of n-alkanes and isoprenoids
are shown in Fig. 6. From internal calibration, the
relative intensities of n-C_{11} to n-C_{33} exhibit a normal
distribution for both coals (Table III). In general for
Crested Butte (Fig. 6b) the distribution pattern is simpler,
when compared with the Grand Mesa (Fig. 6a) in which the
distribution pattern is complex. According to Bartle et al.
(10) distribution of a-alkanes may spread from C_{11} to C_{33}.
Diagenetically, it is possible that the C_{25}-C_{33}
hydrocarbons may be derived principally from higher plants,
C_{12}-C_{20} hydrocarbons may be from an algal source, and the
C_{15}-C_{20} may come from catagenesis. For both samples it
appears that C_{19}-C_{20} is at a maximum. In the Grand Mesa
sample, overlapping of a multiple distribution becomes
apparent (Fig. 7). This multimodal pattern may reflect
geological immaturity in Grand Mesa, since it contains
groupings of overlapped assemblages of molecules. As far
as the predominance of odd-carbon numbered n-alkanes is
concerned, both samples are similar without distinctive
differences. The carbon preference index (CPI) calculated
for Grand Mesa is 1.18 and for Crested Butte 1.10, (Fig.
7a & b).   Similarly the odd-even preference (OEP) based
on C_{25} is 1.08 for Grand Mesa and 0.99 for Crested Butte
(Fig. 7 a & b). The slightly lower values of both
indices for Crested Butte signify that the sample is
more mature. The reason is derived from the fact that

TABLE III   n-Alkanes in the Hexane Fraction.

Carbon No.	Relative Abundance	
	Grand Mesa	Crested Butte
Standard	8	8
11	–	0.5
12	1	3
13	2	4.45
14	6	33
15	20.5	40
16	6.5	68.5
17	11	66
18	11	70
19	15.2	80
20	28.5	65
21	12.5	61
22	10	48
23	7.5	46
24	9.5	40
25	8	35
26	4	30
27	3	25
28	4	20
29	5	14
30	3	9
31	2	8
32	2.5	5
33	2	5

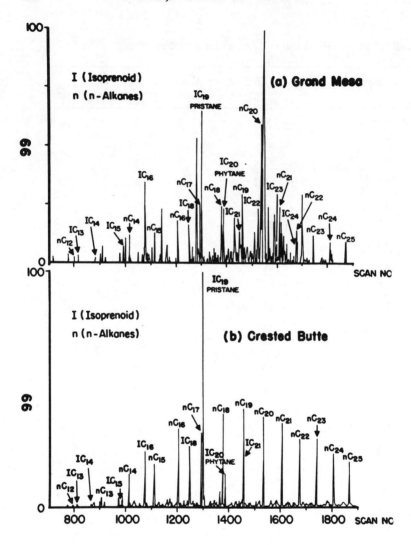

FIGURE 6   The n-alkanes and isoprenoid hydrocarbons of
(a) Grand Mesa and (b) Crested Butte. Notice
the absence of IC_{17} of both samples. The
fragmentograms are from the monitoring of mass
ion m/z in GC/MS.

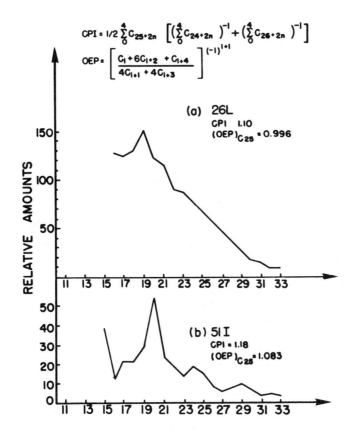

$$CPI = 1/2 \sum_{0}^{4} C_{25+2n} \left[\left(\sum_{0}^{4} C_{24+2n} \right)^{-1} + \left(\sum_{0}^{4} C_{26+2n} \right)^{-1} \right]$$

$$OEP = \left[\frac{C_i + 6C_{i+2} + C_{i+4}}{4C_{i+1} + 4C_{i+3}} \right]^{(-1)^{i+1}}$$

(a) 26L
CPI 1.10
$(OEP)_{C25} = 0.996$

(b) 51 I
CPI = 1.18
$(OEP)_{C25} = 1.083$

RELATIVE AMOUNTS

NUMBER OF CARBON ATOMS OF n-ALKANE

FIGURE 7   The carbon distribution patterns of (a) Crested
            Butte and (b) Grand Mesa together with the
            calculated values of CPI and OEP.

preservation of odd predominance is commonly observed for
biological n-alkanes although it is progressively
obliterated with increasing depth of burial and age. This
conclusion is in accordance with that of White et al. (11).

High rank coals usually do not show predominance of odd over even (12,11), low rank coal usually do (13). Radke et al. (14) correlate changes in n-alkane distribution of coal extracts as a function of increasing maturity, measured by vitrinite reflectance, R_o. For coal with $R_o > 0.95$ there will be no odd preference, and vice versa with coals having $R_o < 0.95$.

Returning to Fig. 6, Grand Mesa centers on (the stongest peak) C_{20}, whereas Crested Butte centers on C_{19}. The isoprenoid IC_{19} (pristane) are prominant in both samples. Isoprenoids IC_{13} to IC_{24} are abundant, except IC_{17}. The pristane/phytane ratio (pr/ph) is 2.8 for Grand Mesa, and 5.4 for Crested Butte. All this suggests that presence of continental organic matter. Apparently the formation of isoprenoids higher than phytane, i.e., IC_{25}, cannot be explained from phytol diagenesis. They may be derived from sesterterpenes (C_{25}) of leaves similar to the genesis of $IC_{15}-IC_{20}$ from phytol. An alternative explanation is from transalkylation. (15). It seems that there is more scrambling from the earlier maturity stage of Grand Mesa than toward a later stage of Crested Butte. For both n-alkanes and isoprenoid, there is a distinctive saw-tooth type distribution for Grand Mesa, but for Crested Butte, a simplified distribution for both types of compounds is observed.

The pr/ph parameter can be an indicator of paleo-environment. Anoxic environments have low pr/ph values (i.e., <<1). Alternating oxic and anoxic conditions may be reflected in pr/ph value of about 1. Wholely oxic conditions produce pr/ph values larger than 1. In studies of a series of coals with various ranking Bartle et al. (10) confirmed the diagenetic process suggested by Brooks

et al. (16) that pr/ph maximumizes at coal carbon contents
of 83-85%. For example, the pr/ph ratio for Turkish lignite
is 1-2 (%C content from 71.1 Seyitomer, to 74.6 Tuncbilek)
and for a Markham Main UK coal (carbon content 82.7%) it is
5.6 (10). These values support our interpretation. The
higher values of Crested Butte may indicate that
maturation may not erase the entire paleoenvironmental
record.

Refering to the aromatic and polynuclear aromatic
hydrocarbons (Table IV), Crested Butte is devoid of the
monoaromatics,but highly peri condensed type systems such
as coronene are abundant. This characteristic is usually
in agreement with that of older shale samples such as
Devonian black shale (17). Monoaromatics, such as cumene,
may be derivatives representative of monoterpenes.

For Grand Mesa, the absence of acenaphthenes and
fluorenes may be significant. In sesquiterpene (Fig. 8)
and diterpene (Fig. 9) chemistry, the biogenic molecules
(B), the transition intermediate molecules (I) and the
thermally altered end products (E) are all different. For
the end products there can be both drastically changed end
products (DE) or mildly altered end products (ME).
Strangely for Grand Mesa, we observed at m/e 183 a strong
peak for cadalene, at 219 a strong peak for retene and also
at 237 a moderately strong simonellite peak. None of these
were found in Crested Butte.. In Crested Butte, we found
both acenaphthene and fluorene; but those were absent in
Grand Mesa. For sesquiterpenes, cadalene is ME and acenaphthene is DE. Similarly for the diterpenes, simonellite is I,
and retene is ME but fluorene is DE. Therefore, the
severity of thermal stress under catagenesis exhibited by
Crested Butte versus the mild conditions of Grand Mesa

TABLE IV   Aromatics and PNAs in Hexane Fraction.

Mass Ions	Compound	
	Grand Mesa	Crested Butte
105	Cumene	-
106	Ethylbenzene p-Xylene + m-Xylene o-Xylene	- - -
128	Naphthalene	Naphthalene
134	1,2,3,4-Tetramethylbenzene	1,2,3,4-Tetramethylbenzene
142	1-Methylnaphthalene 2-Methylnaphthalene	1-Methylnaphthalene 2-Methylnaphthalene
154	Biphenyl -	Biphenyl Acenaphthene
156	2,6-Dimethylnaphthalene	2,6-Dimethylnaphthalene
166	- -	Fluorene 2-Methyfluorene
170	2,3,5-Trimethylnaphthalene	-
178	Phenanthrene Anthracene	Phenanthrene Anthracene
184	-	Dibenzothiophene
202	Fluoroanthene Pyrene	Fluoroanthene Pyrene
216	-	2,3-Benzofluorene
228	Benzo(a)anthracene Chrysene	Benzo(a)anthracene Chrysene
252	Benzo(b)fluoranthene Benzo(k)fluoranthene Benzo(e)pyrene Benzo(a)pyrene Perylene	- - Benzo(e)pyrene Benzo(a)pyrene Perylene
276	Benzo(g,h,i)perylene Anthanthrene	Benzo(g,h,i)perylene Anthanthrene
278	-	1,2,5,6-Dibenzoanthracene
300	-	Coronene

are obvious.

For a series of coals from low to high rank. Gellegos

SESQUITERPENES

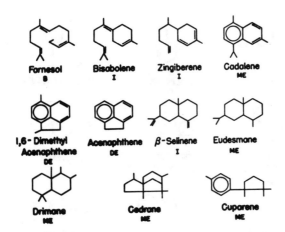

FIGURE 8   Molecular types of sesquiterpenes.   B is biogenic
           molecule, I is intermediate molecule, DE is
           drastically changed end molecule, and ME is
           mildly altered end molecule.   Bisabolene is
           obtained from fornesal (the OH is not shown),
           β-selinene and drimane may came from another
           sources.   1,6-Dimethylacenaphthene may be derived
           from cadalene by condensation.

(18) suggested that cadalene is a sensitive biomarker, since

it can only be detected in the lower rank coals from

lignite, subbituminous coal and bituminous C, and it cannot

be detected in higher rank coals such as bituminous B,

bituminous A, etc.   It should be emphasized here that ME

molecules do not require carbon rearrangement or ring

cleavage of side chain scission which are required from DE

molecules.   ME molecules only require aromatization.   There-

fore, partially aromatized or fully aromatized molecules

such as simonellite, retene and cadalene always appear in

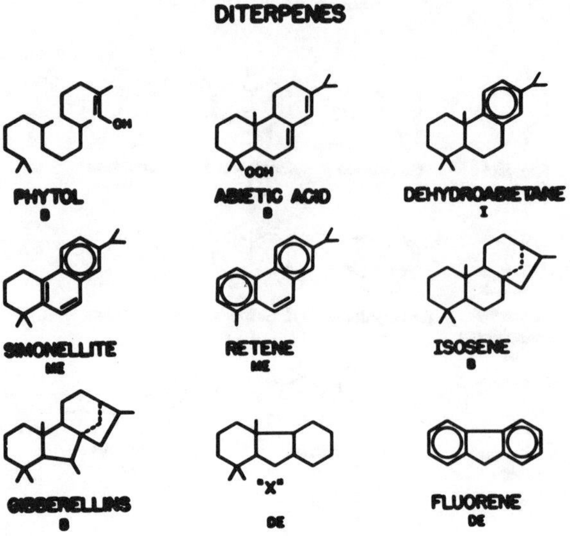

DITERPENES

FIGURE 9   Molecular types of **diterpenes**. The symbols used
           are as Fig. 8.   Fluorene and compound x may be
           derived from giberellins or isosene.

low rank coals. However, for DE molecules, heat and
pressure are required; in this manner some high rank coals
have undergone catagenesis under these conditions. In the
case of Crested Butte, evidence of metamorphic intrusions
is known (2).

For triterpanes, the m/e 101 fragmentograms (Fig. 10)
provide prominent peaks whose identifications are listed in
Table V. For Crested Butte (Fig. 10b) the strongest peaks
are E, $17\alpha(H)$, $21\beta(H)$-30 norhopane, I, $17\alpha(H)$, $21\beta(H)$-
hopane and M, $22-S-17\alpha(H)$, $21\beta(h)$-30 homohopane. Also
Crested Butte contains all both C-22 isomers from C_{31} to
C_{34} in pairs, e.g., MN, QR, TU, YZ. They are all absent
for the Grand Mesa except the 22R isomer of C_{31} (N).

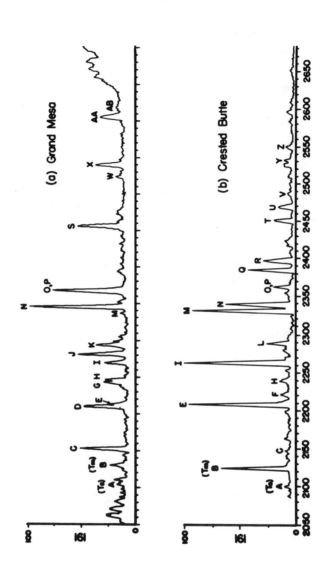

FIGURE 10   Triterpanes of (a) Grand Mesa and (b) Crested Butte as obtained from m/e 191. The capital letters are compounds identified as listed in Table 5.

TABLE V   Triterpanes Identified In Southeastern Uinta
          Region Coals.

Compound	Name	Grand Mesa	Crested Butte
$C_{27}H_{46}$	18α(H)-22,29,30 trisnorhopane (Ts)	A	A
$C_{27}H_{46}$	17α(H)-22,29,30 trisnorhopane (Tm)	B	B
$C_{27}H_{46}$	17β(H)-22,29,30 trisnorhopane	C	C
$C_{29}H_{46}$	Unknown triterpane	D	--
$C_{29}H_{50}$	17α(H),21β(H)-30 norhopane	E	E
$C_{30}H_{52}$	Iupane (C$_{30}$ triterpane)	--	F
--	Unknown (MW 360 ?)	G	--
$C_{29}H_{50}$	17β(H),21α(H)-normoretane	H	H
$C_{30}H_{52}$	17α(H),21β(H)-hopane	I	I
$C_{30}H_{50}$	iso-hop-13(18)-ene	J	--
$C_{29}H_{50}$	17β(H),21β(H)-30 norhopane	K	--
$C_{30}H_{52}$	17β(H),21α(H)-moretane	--	L
$C_{31}H_{54}$	22S-17α(H),21β(H)-30 homohopane	M	M
$C_{31}H_{54}$	22R-17α(H),21β(H)-30 homohopane	N	N
$C_{30}H_{52}$	17β(H),21β(H)-hopane	O	O
$C_{31}H_{54}$	17β(H),21α(H)-homomoretane	P	P
$C_{32}H_{56}$	22S-17α(H),21β(H)-30,31-bishomohopane	--	Q
$C_{32}H_{56}$	22R-17α(H),21β(H)-30,31-bishomohopane	--	R
$C_{31}H_{54}$	17β(H),21β(H)-30 homohopane 20R	S	--
$C_{33}H_{58}$	22S-17α(H),21β(H)-30,31,32-trishomohopane	--	T
$C_{33}H_{58}$	22R-17α(H),21β(H)-30,31,32-trishomohopane	--	U
$C_{32}H_{56}$	17β(H),21α(H)-30,31,32-trishomohopane	--	V
$C_{32}H_{56}$	17β(H),21β(H)-30,31 bishomohopane	W	--
--	Unknown, MW 432 terpenoid	X	--
$C_{34}H_{60}$	22S-17α(H),21β(H)-30,31,32,33-tetrakishomohopane	--	Y
$C_{34}H_{60}$	22R-17α(H),21β(H)-30,31,32,33-tetrakishomohopane	--	Z
--	Unknown, MW 446 terpenoid	AA	--
$C_{33}H_{58}$	17β(H),21β(H)-30,31,32,trishomohopane	AB	--

The contrast in concentration of M exhibited in both
samples is noteworthy.

      For both samples there are no C_{28} hopanes, and also

for Grand Mesa there is no distinct predominance of C_{29}
over C_{30} hopanes; apparently the depositional environment
cannot be only euxinic. Nevertheless, useful information
can be derived based on the stereochemistry of C-17 and
C-21 in hopanes greater than C_{29}. Living organisms always
synthesize β,β-hopanes 17β(H), 21β(H) (and sometimes
subordinate β,α hopanes). The α,β-hopanes have never been
detected in organisms. Therefore the α,β-series detected
are generated by substage isomerization of the β,β-type or
β,α-type. Thus, $\beta,\beta/\alpha,\beta$ is an aging indicator for coals.
In Grand Mesa for every β,β-hopane, there is a corresponding α,β isomer. The ratio $\beta,\beta/\alpha,\beta$ is about 1 for every
isomer pairs. However, in Crested Butte, all but one
exception, the β,β-isomer has been transformed into α,β
-isomer. The exception is β,α-homomoretane, P. Even in
this case the $\beta,\beta/\alpha,\beta$ ratio is less than 0.1. This
information suggests that the Crested Butte is more mature
than Grand Mesa. Ultimate stability of the stereochemistry
increases in the order 17β(H), 21β(H)<17β(H), 21α(H)<17α(H),
21β(H). The α,β hopanes are most stable. This explains
the abundant quantity reflected by peak intensity in E, I
and M of the Crested Butte sample.

Another feature of the triterpane series is the
epimerization of C-22 in hopanes of more than C_{30}. In our
analysis the 22S epimer elutes before 22R epimer in every
case. In the extended hopane series, C_{31}-C_{35}, biogenic 22R
epimers are converted to an equilibrium mixture of 22R and
22S epimers; thus the 22S/22R ratio can be an indicator of
the degree of diagenesis. For Grand Mesa this indicator
is 0.2, while for Crested Butte, the indicator becomes
1.2, signifying that considerable maturation has taken
place for the latter sample. The Tm/Ts ratio is not a

useful indicator for the two samples studied, since for
both samples Ts<Tm. It is possible that the temperature
and pressure experienced in Crested Butte was not severe
enough to cause carbon skeleton migration (e.g., Ts needs
migration of methyl group from Tm). This is consistent
with the observation that no severe demethyled products
were isolated for Crested Butte. Neither sample shows
signs of extensive biodegradation.

At m/e 217, a number of steranes have been detected
(Fig. 11), although the sterane/hopane ratio is low. From
Fig. 11 and Table VI, it appears that C_{29} stigmastanes
predominate over the C_{27} cholestanes. All those points
indicate that the samples are land-derived ediments
(continental organic matter) as suggested by Powell &
McKirdy (19). Epimerization for C-20 in steranes is a good
biomarker indicator, e.g., the 20S/20R in C_{27}-C_{29} steranes.
Increasing maturation usually changes the configuration at
C-20 from a 20R predominance to equal amounts of 20R and
20S. For Crested Butte, all cholestanes and diacholestanes
yield 20R and 20S pairs. In contrast, the Grand Mesa does
not have any cholestanes and diacholestanes, therefore the
20S/20R values cannot be evaluated. For diacholestanes the
value of 20S/20R is 0.3 for Crested Butte. For Grand Mesa,
however, we have isostigastane (ag and ah) for comparison.
Hence 20S/20R values are 0.5 for Grand Mesa and 0.9 for
Crested Butte. The higher the value, the more mature
the sample. In this manner of assessment Crested Butte
appears more mature.

For Grand Mesa, apart from the fact there are no C_{27}
and C_{28} regular steranes and isosteranes, the sample
contains numerous rearranged and aromatized steranes. In
contrast, there are neither severely rearranged nor

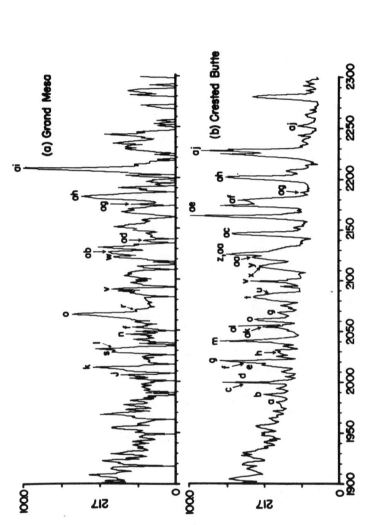

FIGURE 11   Steranes of (a) Grand Mesa and (b) Crested Butte as obtained from m/e 217. The lower case letters indicate the compounds identified and listed in Table 6:

TABLE VI   Steranes Identified in Southeastern Uinta Region Coals.

Compound	Name	Grand Mesa	Crested Butte
- -	C_{28} rearranged steranes	i	- -
$C_{27}H_{48}$	13β,17α-diacholestane (20S)	- -	a
$C_{27}H_{48}$	13β,17α-diacholestane (20R)	- -	b
$C_{26}H_{46}$	C_{26} sterane	- -	c
- -	Monoaromatized steranes	j	- -
$C_{27}H_{48}$	13α,17β-diacholestane (20S)	- -	d
$C_{27}H_{48}$	13α,17β-diacholestane (20R)	- -	e
- -	C_{29} rearranged steranes	k	- -
$C_{28}H_{50}$	24-methyl-13β,17α-diacholestane (20S) or (24S&R)	- -	f
$C_{28}H_{50}$	24-methyl-13β,17α-diacholestane (20S) (isomers in mixture)	- -	g
- -	Monoaromatized steranes	l	- -
$C_{27}H_{48}$	Rearranged C_{27} sterane	- -	h
$C_{28}H_{50}$	24-methyl-13β,17α-diocholestane (20R) or (24S&R)	- -	m
- -	Monoaromatized steranes	n	- -
$C_{29}H_{52}$	24-ethyl-13β,17α-diacholestane (20S or R in mixture)	o	o
- -	C_{29} rearranged steranes (MW 400)	p	- -
$C_{28}H_{50}$	24-methyl-13α,17β-diacholestane (20R)	- -	q
- -	Monoaromatized steranes	r	- -
- -	Monoaromatized steranes	s	- -
$C_{27}H_{48}$	5 (H)-cholestane(5α,14α,17α-cholestane (20R))	- -	t
$C_{29}H_{52}$	24-ethyl-13β,17α-diacholestane (20R)	- -	u
$C_{29}H_{52}$	C_{29} sterane (mixed c̄ monoaromatic sterane)	v	- -
$C_{29}H_{52}$	Rearranged C_{29} sterane	w	- -
$C_{29}H_{52}$	24-ethyl-13α,17β-diacholestane (20S)	- -	x
$C_{28}H_{50}$	24-methyl-5α,14α,17α,cholestane (20S) (5α-ergostane 20S)	- -	y
$C_{29}H_{52}$	24-ethyl-13α,17β-diacholestane (20R)	- -	z
$C_{28}H_{50}$	24-methyl-14β,17β-cholestane (20R,20S in mixture)	- -	aa
- -	Monoaromatized steranes	ab	- -
$C_{28}H_{50}$	24-methyl-5α,14α,17α,cholestane (20R) (5α-ergostane (20R))	- -	ac
- -	Monoaromatized steranes	ad	- -
$C_{29}H_{52}$	24-ethyl-5α,14α,17α-cholestane (20S) (5α-stigmastane 20S)	- -	ae
$C_{29}H_{52}$	24-ethyl-5β,14α,17α-cholestane (20R) (5β-stigmastane 20R)	- -	af
$C_{29}H_{52}$	isostigmastane 20R	af	ag
$C_{29}H_{52}$	isostigmastane 20S	ah	ah

TABLE VI   (cont.)

Compound	Name	Grand Mesa	Crested Butte
$C_{29}H_{52}$	5α(H)-stigmastane (5α,14α,17α-stigmastane 20S or R) [ai]	- -	
$C_{30}H_{54}$	C_{30} steranes	- -	aj
$C_{28}H_{50}$	24-methyl-13α,17β-diacholestane (20S)	- -	ak
$C_{27}H_{48}$	5α,14α,17α-cholestane (20S)	- -	al

aromatized steranes in Crested Butte (Table VI). Since in
general for a triangular diagram (20), C_{27} sterols dominate
in marine organisms, C_{28} sterols indicate lacustrine matter,
and C_{29} sterols are more common in terrestrial plants.
Both samples originate from land plants but Crested Butte
may have been exposed to waters of unknown nature. The
large number of species of aromatic and rearranged steranes
in Grand Mesa sample may be directly related to the mildly
altered source material. Epimerization at the C-14 and
C-17 positions yield isoskeletal steranes of the β,β type
since at equilibrium the more stable β,β-form is favored.
In the more mature Crested Butte sample 24-methyl-14β,17β-
cholestane (aa) is quite abundant. It is essential to point
out that aromatized species under oxidation conditions
(Crested Butte) do not survive.

We have tentatively evaluated the two coals in the
Southeastern Uinta Regions. Geographically, their deposits
are close and their characterization cannot be easily
differentiated by ordinary chemical methods, e.g., their
H/C's are similar, and hence their aromaticity values f_a are
also close. Through the Van Krevelen plot, both coals
belong to type III kerogen. In the present studies, we
have shown some of the simple biomarker parameters (Table
VII) to differentiate the two coals.

TABLE VII   Some Biomarkers for South Eastern Uinta
             Region Coals.

Biomarker parameters	Grand Mesa	Crested Butte
DPEP/Etio	0.1 - 0.2	0.03
CPI	1.18	1.10
OEP	1.08	0.99
Pr/Ph	2.8	5.4
Homohopane 22S/22R	0.2	1.2
Hopane C-17 & C-21, $\beta,\beta/\alpha,\beta$	1	<0.1
Stigmastanes 20S/20R	0.5	0.9

Grand Mesa, a subbituminous A, is lower rank coal when
compared with Crested Butte, a high volatile bituminous B.
The Grand Mesa is mainly a vitrain, whereas the Crested
Butte can be classified as a durain. Usually vitrain is
developed in relatively dry conditions under a highly
anaerobic environment. On the other hand durain is produced
in true wet swamp, where aeration and water are constant
companions. Concerning the rain precipitation,this is even
true today; maybe it also holds during genesis since Late
Cretaceous time. All these facts are in agreement with the
observation of large numbers of aromatic sesqui- and diterpenes and aromatic steranes in Grand Mesa. These species
will not survive in the oxygeneous depositional environment
of Crested Butte. Also, in an environment where water is
plentiful, biodegradation becomes important. We also have

epimerization evidence for advanced maturation of Crested
Butte in agreement with the ranking of these coal samples.

4. CONCLUSIONS

Tentatively the following conclusions can be made:

    1. Biomarker studies support the fact that the Crested
Butte coal is typically of higher rank and more mature,
and the Grand Mesa coal is of lower rank and less mature.

    2. The absence of acenaphthenes and fluorenes and the
presence of cadalene and retene and/or simonellite may be
related to mild catagenesis. For severe catagenesis, this
trend is reversed. As such they can be useful biomarkers.

    3. Both coal samples contain similar biomass source
materials. During diagenesis, Grand Mesa is exposed to an
anaerobic environment with less water contact. Crested
Butte is in an aerobic environment, constantly in contact
with water, and perhaps proceeds with more biodegradation.

    4. Crested Butte coal has experienced some heat and
pressure, but not extensively nor excessively.

    5. Geochemical biomarkers may act as major parameters
for certain given fossil remains (e.g., coals).

ACKNOWLEDGEMENT

    The authors are appreciative of the financial support
from U.S. DOE Contract No. ET-78-G-01-3379 (12162) and
AS03-76-EV10017 and ACS PRF Grant No. 16319-AC5,2. They
also acknowledge the coal samples provided from J.M.
Karprinski of Bethlehem Steel Corporation. Y.D.G. and Y.W.
also want to thank their hosting institutions for the
opportunities to visit USC.

REFERENCES

1. E. R. LANDIS, U.S. Coal Survey Bull., 1072-C, 131-229 (1959).
2. E. C. DAPPLES, Econ. Geol., 34, 369-398, (1939); also 35, 109, (1939).
3. E. W. BAKER, T. F. YEN, J. P. DICKIE, R. E. RHODES and L. F. CLARK, J. Am. Chem. Soc., 89, 3631-3639, (1967).
4. T. F. YEN, L. J. BOUCHER, J. P. DICKIE, E. C. TYNAN and G. B. VANGHAN, J. Inst. Petrol., 55, 87-99, (1969).
5. A. J. G. BARWISE and P. J. D. PARK, Adv. Org. Geochem. 664-674, (1981).
6. J. M. E. QUIRK, G. EGLINTON and J. R. MAXWELL, J. Am. Chem. Soc., 101, 7693-7697, (1979).
7. T. F. YEN and S. R. SILVERMAN, Am. Chem. Soc. Div. Petroleum Chem. Preprints, 14(3), E32-E39, (1969).
8. B. M. DIDYK, Y. I. A. ALTURKI, C. T. PILLINGER and G. EGLINTON, Nature, 256, 563-565, (1975).
9. R. BONNETT and F. CZECHOWSKI, Nature, 283, 465-467, (1980).
10. K. D. BARTLE, D. W. JONES, H. PAKDEL, C. E. SNAPE, A. CALIMLI, A. OLCAY and T. TUGRUL, Nature, 277, 284-287, (1979).
11. C. M. White, T. L. SHULTZ and A. G. SHARKEY, JR., Nature, 268, 620-622, (1977).
12. K. OUCHI and K. IMUTA, Fuel, 42, 445-456, (1963).
13. K. D. BARTLE, T. G. MARTIN and D. F. WILLIAMS, Fuel, 54, 226-235, (1975).
14. M. RADKE, R. G. SCHAEFER, D. LEYTHAEUSER and M. TEICHMULLER, Geochim. Cosmochim. Acta, 44, 1787-1800, (1980).
15. T. F. YEN, Genesis and Degradation of Petroleum Hydrocarbons in Marine Environments, Marine Chemistry in the Coastal Environment, edited by T. M. Church (Am. Chem. Soc. Symposium Ser. No18, 1975), pp 231-266.
16. J. D. BROOKS, K. GOULD and J. W. SMITH, Nature, 222, 257, (1969).
17. T. F. YEN, J. I. S. TANG, D. K. YOUNG and E. CHOW, Second Eastern Gas Shale Symposium, Vol. 1, 1978, pp. 330-338.
18. E. J. GALLEGOS, Am. Chem. Soc., Petroleum Chem. Preprints, 22, 604-619, (1977).
19. T. J. POWELL and D. M. MCKIRDY, Am. Assoc. Pet. Geol. Bull., 59, 1176-1197, (1975).
20. A. S. MACKENZIE, Adv. Petroleum Geochemistry, 1, 115-214, (1984).

21. JERRY HAN and MELVIN CALVIN, Geochim. Cosmochim. Acta,
     33, 733-742, (1969).

BIOMARKER COMPARISONS OF MICHIGAN BASIN OILS

K.W. DUNHAM and P.A. MEYERS
Department of Atmospheric and Oceanic Science
The University of Michigan
Ann Arbor, Michigan 48109-2143

J. RULLKÖTTER
Institut für Erdöl und Organische Geochemie
Kernforschungsanglage Jülich GmbH
5170 Jülich
Federal Republic of Germany

Petroleum samples from the Michigan Basin have been
analyzed by capillary gas chromatography, combined gas
chromatography-mass spectrometry, and carbon isotope
mass spectrometry to compare their geochemical characteristics and to detect evidence of their origins.
Oils from the Ordovician-age Trenton and Devonian-age
Dundee formations have similar, mature biomarker
characters. Both contain distributions of n-alkanes,
steranes, and hopanes typical of oils generated from
clay-bearing source rocks containing marine organic
matter. Because they are reservoired in carbonate
rocks having relatively low geothermal temperatures,
the Trenton and Dundee oils probably were generated
deeper in the Basin and migrated to their present
locations. Smaller concentrations of steranes in the
Dundee oils suggest somewhat lower maturity and are
evidence of earlier generation and migration away from
their source beds. Silurian-age Niagaran reefs contain
oils having hydrocarbon patterns different from the
Trenton and Dundee Oils. They are typical of thermally
mature oils from carbonate source-rocks and appear to
contain major contributions from land-derived organic
matter.

1. INTRODUCTION

A variety of geochemical measurements have been used to compare oils to each other and to the organic matter contained in possible source rocks. Some of the goals of such comparisons have been to determine the relationships between petroleum and its precursor organic matter, to ascertain the effects of migration upon oil composition, to reconstruct migrational routes, to define the types of geochemical changes that accompany thermal maturation first within the source rock and later in the reservoir rock, and to delineate degradational modifications of an oil's character brought about by microbes, water washing, or other processes. A few examples of geochemical measurements include carbon isotopes[1], kerogen pyrolysis[2], and acyclic and cyclic aliphatic hydrocarbons[3,4,5]. Such measurements have been determined in the Michigan Basin.

The Michigan Basin is a Paleozoic intracratonic basin. The rocks are typical of such a basin; they are thick, lithified deposits of gently dipping shallow water sediments. Although some Precambrian and Mesozoic strata exist, nearly all of the rocks in the Michigan Basin are Ordovician, Silurian, and Devonian in age. The basin has been filled with predominantly carbonates and very few clastic deposits. No evidence exists for major tectonism, therefore, uplift and erosion have been minimal.

The past geothermal gradient remains an important question as it is directly related to petroleum production. The present thermal gradient is about $22^{\circ}C/km$, obtained from a deep borehole near the center of the basin[6]. Nunn and coworkers[7] developed a petroleum generation model limiting petroleum production to Ordovician strata and only the most

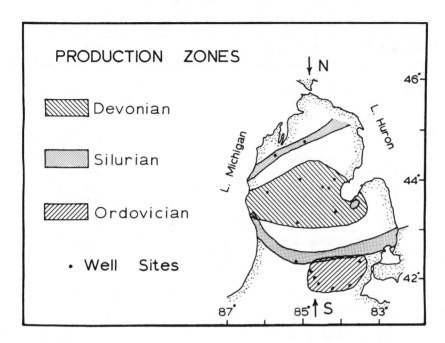

FIGURE 1   Hydrocarbon production zones in the Michigan basin

deeply buried Silurian strata.   In contrast, Cercone[8] concludes from the maturity of kerogen that thermal gradients were once as high as 45°C/km and that as much as 1 km of Carboniferous strata were eroded, permitting oil generation to occur even in Devonian rocks during the Paleozoic.

Hydrocarbon production in the Michigan Basin is divided into four geographic zones, shown in Figure 1.   Ordovician production is in the southeast corner of the basin, Silurian production is from a northern and southern reef trend, and Devonian production is located throughout the center of the basin.

TABLE 1   Michigan Basin oils examined in this study

Reservoir Age and Formation	Field
Mississippian	
Stray Sandstone	Clare
Berea Sandstone	Saginaw
Devonian	
Traverse Limestone	Peacock
Dundee Limestone	N. Adams
Dundee Limestone	Bentley
Dundee Limestone	Crystal
Dundee Limestone	Oil Springs
Detroit River Sour	S. Buckeye
Silurian	
Niagara Limestone	Blue Lake
Niagara Limestone	Grant 26
Niagara Limestone	Pennfield
Ordovician	
Trenton Limestone	Deerfield
Trenton Limestone	Blissfield
Trenton Limestone	Albion—Scipio
Black River Limestone	Albion—Scipio
Trenton—Black River Limestone	Hannover

In this study, 16 Michigan Basin oils were examined (Table 1). The work here supports the earlier findings of Vogler and coworkers[9] between similarities in the oils reservoired in the Trenton Limestone (Ordovician) and Dundee Limestone (Devonian). Other results suggest the possibility of multiple sources for non—Dundee Devonian oils. In this paper, when a geologic age is used in association with an oil, it refers to the age of the reservoir strata.

2. ANALYSIS

2.1. Capillary Gas Chromatography

Whole oil samples were analyzed employing splitless injection on a Hewlett-Packard 5830 gas chromatograph fitted with a flame ionization detector and a 20 m x 0.3 mm i.d. SE54 fused silica capillary column. Hydrogen was used as the carrier gas. Retention times were used to identify individual n-alkanes and isoprenoids. Standards were injected prior to actual analysis to provide retention time indices. After samples were injected at 120°C, the column oven was heated at 6°C/min to 270°C with a 20 minute hold for a total run time of 45 minutes.

2.2. Combined Gas Chromatography-Mass Spectrometry

The volatile components of the oils were removed by rotary evaporation at 70°C under reduced pressure until a constant weight was reached. The C_{15+} portion of the oils was mixed with benzene (5-10 ml) and the benzene insoluble portion determined by filtration. The asphaltenes were precipitated by the addition of n-pentane (50 ml). The aliphatic and aromatic hydrocarbon fractions were separated from the deasphaltened oils using medium pressure liquid chromatography.

Biological marker distributions were obtained by gas chromatography/mass spectrometry of the saturated and aromatic hydrocarbon fractions, respectively, using a Carlo Erba Fractovap model 4160 gas chromatograph fitted with a 25 m X 0.3 mm i.d. SE54 glass capillary column and connected via an open-split coupling to a Kratos AEI MS 3074 mass spectrometer. After splitless injection at 80°C the temperature of the gas chromatography oven was raised to

100°C and programmed at a rate of 4°C/min to 300°C. The mass spectrometer was operated at 70 eV, and the data were collected in the multiple peak monitoring mode using a DS 50S on-line data system (Kratos). A total of six ions characteristic of sterane and terpane fragmentation (m/z 177, 191, 217, 218, 231, 259), were recorded in the saturated hydrocarbon fraction. Dwell times were 200 msec and 250 msec, respectively, resulting in a cycle time of about 2.5 sec/scan. The compounds were identified in their key fragmentograms based on relative retention times and comparison with published data.

2.3. Oil Fractionation and Carbon Isotope Mass Spectrometry

Whole oils were separated into three fractions by column chromatography. The columns were packed with alumina over silica gel. Both absorbants were 5% deactivated. The oils were adsorbed onto alumina and then placed on top of the prepared column. The column was then eluted sequentially with the following; 10 ml of petroleum ether, 10 ml of 85:15 P.E./benzene, and 20 ml of methylene chloride. Due to the increasing polarity in the solvents, the fractions thus obtained from each elution were: aliphatic hydrocarbons, aromatic hydrocarbons, and a combined resin/asphaltene fraction, respectively.

Whole oils and their fractions were combusted in evacuated sealed tubes filled with cupric oxide at 550°C for 12 hours. Stable carbon isotope ratios of the organic carbon content of the oils and their fractions were determined using a VG Micromass 602 mass spectrometer calibrated with NBS-20 (carbonate) and NBS-22 (petroleum) standards. Data are corrected for ^{17}O and are presented in terms of the PDB standard.

2.4 Kerogen Pyrolysis

Eight potential source rocks in the Michigan Basin were subjected to kerogen pyrolysis. Samples were analyzed using a Girdel Rock-Eval II. Operation and interpretation follows that of Espitalié and others[2].

3. RESULTS

Capillary gas chromatography of whole oil samples reveals many similarities between Ordovician and some Devonian oils and dissimilar characteristics in Silurian oils. Chromatograms of three representative oils are shown in Figure 2. These capillary gas chromatography results are similar to previous studies of Vogler and coworkers[9] and Nunn and coworkers[7]. Trenton and Dundee oils have similar distributions dominated by short chain-length n-alkanes. Niagaran oils, however, have a very broad n-alkane distribution. Ordovician and most Devonian oils have a similar Carbon Preference Index (CPI) ca. 1.3, whereas, Silurian oils have a CPI ca. 1.0 or less[9,10]. Concentrations of the isoprenoid hydrocarbons, pristane and phytane, are relatively low in the Trenton and Dundee oils compared to the Niagaran oil. In the Niagaran oil, phytane is more abundant than pristane. In the Trenton and Dundee oils, however, pristane is more abundant.

The use of cyclic hydrocarbons, particularly mass fragments m/z 191 and m/z 217, has proven useful in oil correlation and maturation studies[3,4,5,11]. A brief discussion of the occurrence of triterpanes and steranes in the Michigan Basin follows; details are given by Rull-

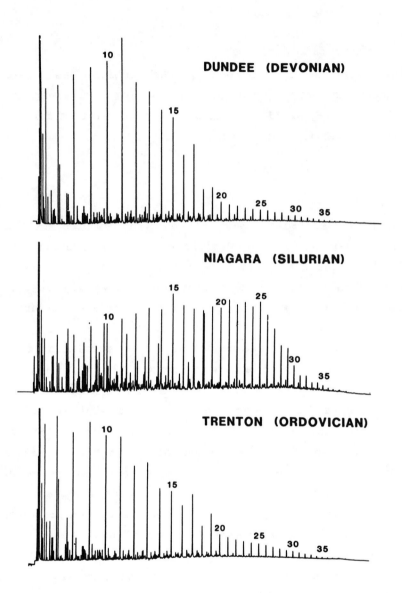

FIGURE 2   Capillary gas chromatographs of three whole oils from the Michigan Basin

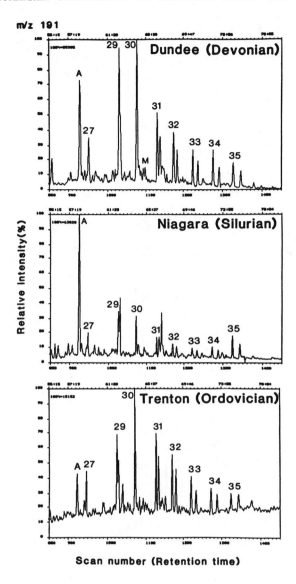

FIGURE 3   Partial triterpane distributions of representative oil types found in the Michigan Basin. Carbon
numbers of 17α(H), 21β(H)-hopanes are indicated.
A=18α(H)-trisnorneohopane; M=unidentified moretanes

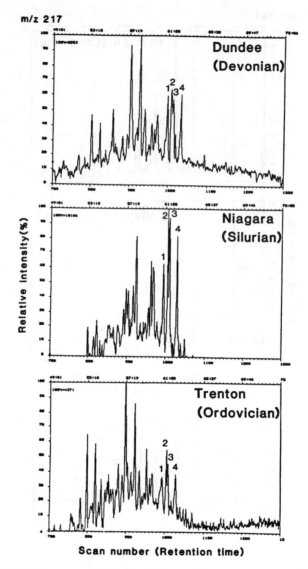

FIGURE 4   Sterane distributions of representative oil types from the Michigan Basin.   Numbered peaks are diastereomers of 24—ethyl—cholestane: 1=14α(H), 17α(H), 20S;   2=14β(H),   17β(H),   20R;   3=14β(H),   17β(H),   20S; 4=14α(H), 17α(H), 20R.

kötter[12]. Distributions of triterpanes and steranes for three oils are shown in Figures 3 and 4, respectively. The Trenton and Dundee oils have similar triterpane and sterane distributions typical of mature oils. Triterpanes are slightly less mature in the Dundee oil. Both oils contain abundant diasteranes. In contrast, the Niagaran oil has low diasterane concentrations and a sterane distribution not fully mature. Furthermore, the ratio of $18\alpha(H)$ to $17\alpha(H)$ trisnorhopanes indicates that Niagaran oils from the northern reef trend are more mature than oils from the southern reef reservoirs[12].

Carbon isotopic compositions have proven useful in oil to oil, migration, and degradation studies[1,13]. The average carbon isotopic pattern of the whole oil and three sub-

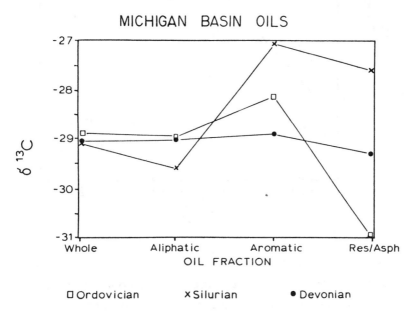

FIGURE 5   Average isotopic values for Ordovician, Silurian, and Devonian oils

fractions is shown in Figure 5. The whole oil isotope value
is similar for all of the oils, averaging −29.0‰ relative to
the PDB. The isotopic pattern of the aliphatic, aromatic,
and resin/asphaltene fractions show no relationship to the
different oils. Figure 6 shows the isotopic relationship
between the aromatic vs. aliphatic fraction for 15 oils.

MICHIGAN BASIN OILS

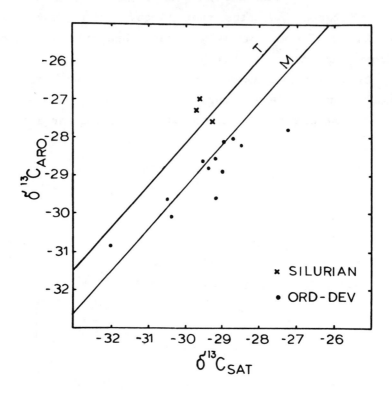

FIGURE 6   Isotopic cross−plot of aromatic vs. saturate
oil fractions.   (T)represents   oils generated from
terrigenous organic matter and   (M)represents oils
generated from marine type organic matter.[1]

The Silurian oils cluster tightly along the terrigenous organic matter line. No difference was observed between the northern and southern reef oils. The Ordovician and

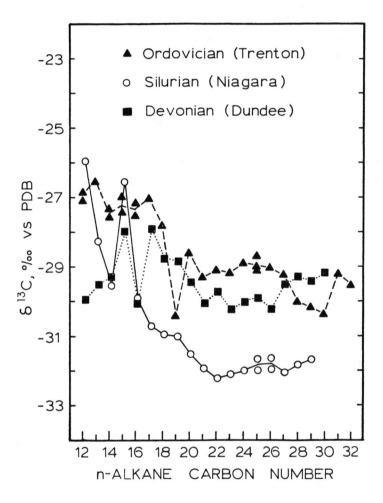

FIGURE 7   Isotopic composition of n-Alkanes for three Michigan Basin oils (from Vogler and coworkers[9])

Devonian oils are distinctly different. They plot along the marine organic matter line. The carbon isotope compositions of individual n-alkanes for three oils are shown in Figure 7. Alkanes above C_{20} are depleted in ^{13}C relative to the shorter chain-lengths. The Niagaran oil is extremely depleted in ^{13}C in the longer "waxy" chain-lengths. The Trenton and Dundee oils are very similar to one another and different from the Niagaran oil.

Organic matter pyrolysis studies can reveal the overall maturity and type of organic matter in source rocks[2]. Potential source rocks in the Michigan Basin were subjected to Rock-Eval analysis. Results are shown in Figures 8 and 9. The low hydrogen index valves (HI's) associated with these samples suggests either very mature source rocks or relatively poor source rocks. All of the samples analyzed had T_{max} values within the oil producing zone.

4. DISCUSSION

4.1 Silurian Oils

It is clear the Silurian oils form a unique group in the Michigan Basin. Since all of the Silurian oils have a broad n-alkane distribution, low CPI's, dominant phytane compared to pristane, and a high percentage of sulfur, the geochemical data suggests a "common" carbonate source rock. This is in agreement with Gardner and Bray[14], who suggest the Salina A-1 carbonates as the principle source rock for the Silurian reef oils. Distribution of steranes also favors a carbonate source rock[12]. The Silurian oils all contain abundant n-alkanes longer than C_{19}. These longer waxy hydrocarbons can be attributed to input from land plants.

MICHIGAN BASIN

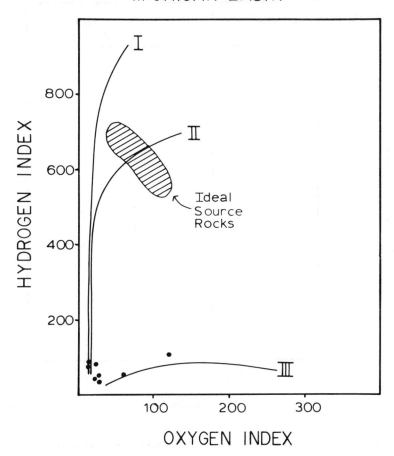

FIGURE 8   Modified van Krevelen plot of Rock Eval
results from potential source rocks

Vascular land plants evolved during the Silurian[15], and the
isotope contents of the aliphatic and aromatic fractions of
the Silurian oils suggest that land–derived organic matter
contributed significantly to these oils[1]. The isotopic
pattern of the n–alkanes[9] also suggests the same interpre–

MICHIGAN BASIN ROCKS

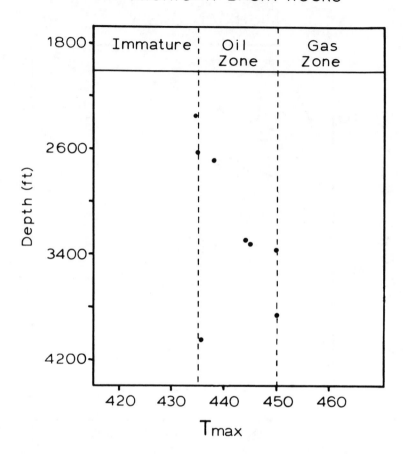

FIGURE 9 T_{max} results from Rock Eval analysis of
potential source rocks

tation since terrestrial material is depleted in [13]C
relative to marine organic material[16].

Subtle geochemical differences allow for further sub-
division into a northern Silurian reef trend and a southern
Silurian reef trend[12,14]. Distributions of steranes and

triterpanes suggest a lower level of thermal maturity for the southern reservoirs. Northern and southern reservoirs also differ in the percentage of sulfur and API gravity[10] (Table 2). In general, the Silurian oils are very similar

TABLE 2   Averaged characteristics of Silurian Oils, from Gardner and Bray[14]

	Depth	% Sulfur	^{O}API
Northern Trend (18 oils)	4,991'	0.43 ± 0.18	42.09 ± 2.84
Southern Trend (16 oils	2,986'	1.16 ± 0.87	35.22 ± 5.07

and suggest a common carbonate source rock in the center of the basin. A common basinward source rock would allow for the differential entrapment of oil and gas in the reef trends[17]. The subtle geochemical differences can be explained by the differences in sub-surface depth in the two reservoirs. Assuming a 22°C/km thermal gradient, the northern reefs are 15°C warmer. This may be sufficient to produce the observed sterane and triterpane maturity differences.[12] The shallower southern reservoirs may be susceptible to degradation by microbes and water washing. Water washing would remove more polar or lighter hydrocarbons. This could account for the API gravities differences observed. Often, water washing is associated with bacterial degradation[18]. Bacterial degradation may account for the increased sulfur content in the southern reservoirs.

4.2. Ordovician and Devonian Oils

The second geochemically distinct group is the Trenton and Dundee oils. Trenton and Dundee oils have n-alkane distributions similar to other oils labeled marine-type by Philippi[19] and are indicative of marine algal organic matter. A marine-type organic matter precursor for the Ordovician and Devonian oils is also suggested by the isotopic relationship between the aliphatic and aromatic hydrocarbons (Figure 6). Abundant C_{27} and C_{29} diasterane distributions suggest a clastic source rock for the oils[12,20]. With the majority of the basin strata consisting of carbonate rocks[21], ideal clastic source rocks are limited. The late-Devonian Antrim shale is the only organic-rich clastic deposit in sufficient quantities to be considered a viable source rock. Even by using the highest proposed thermal gradient it is unlikely that the Antrim shale is of sufficient thermal maturity. The source rock for the Trenton and Dundee oils is most likely a clay bearing or "dirty" carbonate. Due to spatial considerations the oils probably have one common basinward source rock.

Other Devonian oils, however, may have multiple sources and complicated migrational histories. Pruitt[22] reported geochemical evidence suggesting that Detroit River oils are a mixture of Ordovician and Silurian oils. Other geochemical evidence suggests that the Traverse oils form another distinct Devonian oil group[10]. Another oil group, in the Michigan Basin, is the Mississippian age Stray and Berea Sandstone oils. Isotopic composition of the aliphatic and aromatic hydrocarbon fractions indicate a marine organic matter origin, similar to Ordovician and Devonian oils.

Both oils have aliphatic and aromatic fractions more depleted in ^{13}C than other Michigan Basin oils. This depletion may indicate another case of early migration away from the same source rock[18].

4.3. Maturity

The production of hydrocarbons from organic matter is a complex relationship between time, temperature, and the type of organic matter. Models for basin maturity[18,23,24] can give significantly different results in older sedimentary basins. The overall relationship between temperature and observed organic maturity is a subject of disagreement[25]. It has been suggested numerous times that organic matter in the Paleozoic may yield anomalous maturity values. This paper cannot support this idea directly, however, the Rock-Eval data suggests unusually mature organic matter compared to bio-marker results. Due to the positioning of the samples in Figure 10, Devonian strata appear to be of sufficient maturity to produce hydrocarbons. This is in direct conflict with the maturity model of Nunn and coworkers[7], which we favor. Another unproven factor not considered in maturation models is the role of sulfur in petroleum evolution. It has been suggested that sulfur lowers the total energy required to produce petroleum from organic matter[26]. The suggested source rock for the Silurian oils is the Salina A-1 carbonate, which was deposited in a hypersaline environment[14]. The observed maturity in the Silurian, not predicted by thermal models, may be a result of abundant sulfur available during the Silurian.

It has been postulated that evaporitic environments can be prolific producers of organic matter[27]. Although bio-

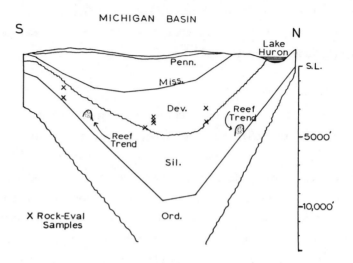

FIGURE 10   Stratigraphic position of Rock Eval samples

logical speciation decreases in hypersaline environments,
overall productivity increases. The evaporitic depositional
conditions, coupled with the relatively old age of the
Michigan Basin, may contribute to many unusual organic
geochemical results.

5. CONCLUSIONS

The Ordovician-age Trenton and the Devonian-age Dundee oils
form a geochemically distinct group in the Michigan Basin.
They are typical of oils generated from clay-bearing source
rocks containing marine organic matter. Subtle geochemical
evidence in the Dundee oil suggests an earlier generation
and migration away from a common source rock. The Silurian-
age Niagaran reef oil is distinctly different from other
oils in the Michigan Basin and is typical of oils generated

from carbonate source rocks. Geochemical evidence suggests that the oil contains major contributions from land-derived organic matter. Small geochemical differences between northern and southern reef oils exist and are the result of post-reservoir alteration.

Other Devonian-age oils form separate geochemically distinct groups. These oils may represent mixtures of oils, oils transformed by some combination of migration and maturation, or oils generated from separate source rocks.

REFERENCES

1. Z. SOFER, AAPG Bull., 68, 31-49, (1984).
2. J. ESPITALIE, M. MADEC, B. TISSOT, J.J. MENNIG, and P. LEPLAT, Offshore Tech. Conf., 9, 439-444, (1977).
3. W.K. SEIFERT and J.M. MOLDOWAN, Geochim. Cosmochim. Acta, 42, 77-95, (1978).
4. W.K. SEIFERT and J.M. MOLDOWAN, Geochim. Cosmochim Acta, 45, 783-794, (1981).
5. A.S. MACKENZIE, V. DISKO, and J. RULLKÖTTER, Org. Geo-chem., 5, 57-63, (1983).
6. N.H. SLEEP and L.L. SLOSS, J. Geophys. Res., 83, 5815-5819, (1978).
7. J.A. NUNN, H.H. SLEEP, and W.E. MOORE, AAPG Bull., 68, 296-315, (1984).
8. K.R. CERCONE, AAPG Bull., 68, 130-136, (1984).
9. E.A. VOGLER, P.A. MEYERS, and W.E. MOORE, Geochim. Cosmochim. Acta, 45, 2287-2293, (1981).
10. H.A. ILLICH and P.L. GRIZZLE, Geochim. Cosmochim. Acta, 47, 1151-1155, (1983).
11. R.P. PHILP, Geochim. Cosmochim. Acta, 47, 267-275, (1983).
12. J. RULLKÖTTER, P.A. MEYERS, R.G. SCHÄFFER, and K.W. DUNHAM, Adv. Org. Geochim., in press, (1986).
13. W.J. STAHL, Geochim. Cosmochim. Acta, 44, 1903-1907, (1980).
14. W.C. GARDNER and E.E. BRAY, Oils and Source Rocks of Niagaran Reefs (Silurian) in the Michigan Basin, Petroleum Geochemistry and Source Rock Potential of Carbonate Rocks, edited by J.G. Palacas, (Am. Assoc. Petrol. Geol., Tulsa, 1984), 33-44.

15. J. GRAY and A.J. BOVCOT, Geol., 6, 489–492, (1978).
16. W.M. SACKETT, Mar. Geol., 2, 173–185, (1964).
17. D. GILL, AAPG Bull., 4, 608–620, (1979).
18. B.P. TISSOT and D.H. WELTE, Petroleum Formation and Occurrence (Springer-Verlag, New York, 1984).
19. G.T. PHILIPPI, Geochim. Cosmochim. Acta, 38, 947–966, (1974).
20. O. SIESKIND, G. JOLY, and P. ALBRECHT, Geochim. Cosmochim. Acta, 43, 1675–1679 (1979).
21. J. DORR and D. ESCHMAN, Geology of Michigan (Univ. of Mich., Ann Arbor, 1970).
22. J.D. PRUITT, Geochim. Cosmochim. Acta, 47, 1159–1161, (1983).
23. J. CONNAN, AAPG Bull., 58, 2516–2521, (1974).
24. D.W. WAPLES, AAPG Bull., 64, 916–926, (1980).
25. I. LERCHE, R.F. YARZAB, and C. KENDALL, AAPG Bull., 68, 1704–1717, (1984).
26. J. RULLKÖTTER, B. SPIRO, and A. NISSENBAUM, Geochim. Cosmochim. Acta, 49, 1357–1370 (1985).
27. D.W. KIRKLAND and R. EVANS, AAPG Bull., 65, 181–190 (1981).

RETENE AND PIMANTHRENE FROM CONTINENTAL SOURCE ROCKS IN CHINA

ZHONG DI CHENG

Institute of Petroleum Exploration and Development
P.O. Box 910, Beijing
People's Republic of China

Retene (1-methyl-7-isopropylphenanthrene), pimanthrene (1,7-dimethyl-phenanthrene) and 1-methyl-7-isopropyl-1,2,3,4-tetrahydrophenanthrene are regarded as representative materials of higher plant lipids and universally exist in the aromatic fraction of continental rock extracts. These aromatic biomarkers are identified by GC/MS in this paper and are used both as a maturity parameter and as indicators of organic matter type. There relative abundance decreases with increasing depth in immature sedimentary rocks. Remarkable changes have been observed at the maturity threshold. The type of organic matter can be recognized according to the ratios of pimanthrene/dimethylphenanthrene, pimanthrene/all alkylphenanthrenes retene/all alkylphenanthrenes. Humics have high ratios although they are lower in mature rock. Sapropelics have very low ratios, but are somewhat higher in the immature rock.

1. INTRODUCTION

The occurrence of complex structures of polycyclic aromatic hydrocarbons (PAH) and their alkyl series from continental source rocks in China has been documented by gas chromatography-mass spectrometry (GC/MS) analysis, and the geochemical significance of these compounds has been discussed.[1] The various carbon numbers of the alkyl side chains and

alkylcycloalkane aromatic hydrocarbons of the formula C_nH_{2n-p}
(p:6,8,10,---,28) have been detected. The percentage of
naphtheno-aromatics is high in immature rocks, but alkylnaphthalenes and alkylphenanthrenes are more concentrated
in mature rocks. Using information obtained from GC analysis,
the author investigates some new geochemical parameters
related to maturity of organic matter and to the depositional
environments of source rocks.

The analyses of PAH in rocks by gas chromatography showed
that alkylnaphthalenes and alkylphenanthrenes are principal
compounds of the aromatic fraction. According to some publications[2,3,4] molecular orbital theory indicates that aromatics
have different thermodynamic stabilities depending upon the
position substituted. The ratios of β-isomers to α-isomers
which depends upon the relative stability of various methyl
substituents in aromatic rings can show the thermal evolution
character of PAH. It is worth noticing that tetrahydroretene
(1-methyl-7-isopropyl-1,2,3,4-tetrahydrophenanthrene), retene
(1-methyl-7-isopropylphenanthrene) and pimanthrene (1,7-dimethyl
phenanthrene), as well as alkylnaphthalenes (e.g. 1,6-dimethyl-
4-isopropylnaphthalene and 1-methyl-7-isopropylnaphthalene)
derived from resin of higher plants, exist universally in the
aromatic fractions of continental source rocks. They are very
significant in geochemistry. It is known that a high abundance of retene and tetrahydroretene exist in the aromatic
fractions of Early Cretaceous source rocks from the Erlian
Basin in Northern China.[4] The Eogene source rocks of the
Beibu Gulf, South Sea, are also rich in retene. Pimanthrene
concentrations are high in the Neogene rocks of the East Sea.
Retene appears in most continental source rocks. It is shown
that thermal maturation and catalysis lead to a generation of
compounds with the abietic carbon skeleton. Typically, con-

tinental humic source rocks are characterized by a high abundance of pimanthrene. The fact that retene and pimanthrene
occur in high abundance in the continental source rocks of
China supports the theory[5,6] that retene can be transformed
from abietic acid and pimanthrene from pimeric acid through
disproportionation.

   In this paper, the relative abundance of the aromatic
constructions shows gradual changes in increasing depth, and
the changes of aromatic contents in different types of rocks
are calculated. The data indicate that the aromatics can be
used as indicators of maturity for continental source rocks
and also as parameters for the classification of organic
matter types.

2. EXPERIMENTAL

   Sample preparation
Rock samples were ground (<150 μm) and Soxhlet-extracted with
chloroform. The extracts were concentrated at room temperature and the chloroform extracts dissolved in petroleum
ehter to remove asphaltenes. The soluble portion was analyzed
by column chromatography. A silica gel and aluminium oxide
(6:4) column was used and the packing material was 200 times
heavier than the sample. The saturated hydrocarbon fraction
was eluted with petroleum ehter, and the aromatic fraction
was eluted with benzene, then concentrated into small tubes
for GC analysis.

   Gas Chromatography
The aromatic fraction was dissolved in hexane and detection
was achieved with a P-E Sigma 2 Gas Chromatography equipped
with a Gerstel inlet splitter, an all-glass outlet splitter,
flame ionization detector (FID) and flame photometric detector

(FPD 394 nm filter). Separation was made on a glass capillary column of 25 m x 0.24 mm I.D. coated with SE-52 silicone gum phase. The oven temperature was programmed from 120°C at 4°C/ min, injector temperature 260°C, detector temperature 300°C. The carrier gas was nitrogen with a split ratio of 1:20 and flow rate of 30 ml/min.

A short column packed with silanized silica gel was lo-cated between the injector and the capillary column to filter non-gaseous and polar material. This filtering column must be cleaned frequently. Splitter linearity and detector sensitivity were frequently checked by running test standard mixtures. The quantitation of indi.idual PAH was performed by comparison with relevant peak areas obtained from a Sigma 10 Intergrator.

GC-MS Analysis

A Finnigan GC-MS system, Model 4021C with electron impact ion source was used for MS identifications at the ionization voltage of 70 ev. The temperature of the ionization source was 250°C. A flexible fused silica capillary column (35 m x 0.26 mm) coated with SE-52 was used. The GC oven temper-ature was 260°C, and the carrier gas was helium.

3. SAMPLES

A total of forty-two source rock samples were studied. Table 1 shows the location of samples and the general description of the rocks. The stratigraphic sections of two wells (Ai and Sai) in the Erlian Basin in the north of China are taken as the main examples. The two wells are situated respectively in the center of the E-he-bao-li-ge and Sai-han -ta-la depression of the basin. The stratigraphic era is from Late Jurassic to Early Cretaceous as shown in Table 2. The source rocks from the wells are mainly argillaceous rocks

Table 1   General description of source rock samples

Sample No.	Location	Geological Era	Depth(m)	Organic Matter type
Ai 1–17	Well Ai of Erlian	Early Cretaceous	450–2000	Sapropel–humic or humic
Sai 1–11	Well Sai of Erlian	Early Cretaceous	1400–2300	Sapropel–humic or humic
Na 1–8	Beibu Gulf	Late Tertiary	1500–2700	Humic-sapropel or sapropel
Du 1–3	East Sea	Early Tertiary	2100–2600	Humic
Li 1–3	Chang qing	Tertiary	1180–1600	Humic-sapropel or sapropel

Table 2   The stratigraphic era of wells Ai and Sai

Era	Early Cretaceous		Late Jurassic	
Formation	Upper coarse interval	Middle fine interval	Lower coarse interval	
Well No.	(m)	(m)	(m)	(m)
Ai	----	Ground–1396.5	1396.5–2000	Under 2000
Sai	Ground–1720	1720–2300	Under 2300	----

deposited in a lacustrine environment.

Analysis of kerogen elements by IR spectroscope, scanning electron-microscope, Rock-eval, etc. show that some formations of Ba-yan-hao and Ba-ta-la-hu of Early Cretaceous in the depressions possess oil generation potential. Unfortunately, though, in the upper coarse interval and the upper part of the middle fine interval of Ba-yan-huo

formation rocks, the organic matter is poor in oil production since coal-like materials are its main constituents.
As for the lower part of the middle fine interval and lower
coarse interval, the organic matter is a mixture of types.
The organic matter of the middle fine interval has a lower
degree of thermal evolution and the vitrinite reflectance
values range from 0.5 to 1.0.

4. RESULTS AND DISCUSSION

4.1. Identification of Peaks

Samples of aromatics from source rocks have been analyzed
by GC. Fig. 1 shows PAH GC chromatograms of an immature
rock extract, aromatic fraction, at 1504m and a mature rock
at 2158m of Sai well in the Erlian basin in the north of
China. Fig. 2 shows PAH GC chromatograms of a typical
humic source rock from the East Sea (top fig.) and a
sapropel type from Beibu Gulf in the South Sea (bottom fig.).
Table 3 shows the identification of chromatogram peaks.

TABLE 3   Identification of peaks

Peak No.	Compound	Mol.
1	Naphthalene	128
2	β -Methylnaphthalene	142
3	α -Methylnaphthalene	142
4	Biphenyl	154
5	β -Ethylnaphthalene	156
6	α -Ethylnaphthalene	156
7	2,6-Dimethylnaphthalene	156
8	2,7-Dimethylnaphthalene	156
9	1,7-Dimethylnaphthalene	156
	1,3-Dimethylnaphthalene	156
10	1,6-Dimethylnaphthalene	156
11	1,4-Dimethylnaphthalene	156
	2,3-Dimethylnaphthalene	156

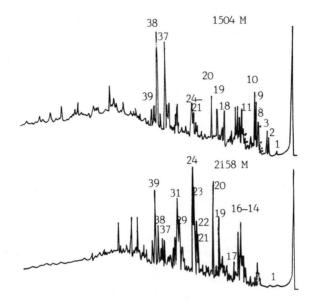

FIGURE 1. PAH GC Chromatograms of immature 1504 M and
mature 2158 M source rocks

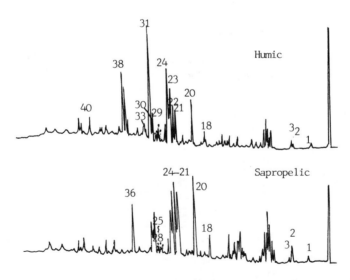

FIGURE 2. PAH GC Chromatograms of a humic source rock
in the East Sea (top) and a sapropelic
source rock in the South Sea (bottom)

12	1,5-Dimethylnaphthalene		156
13	1,2-Dimethylnaphthalene		156
14	1,8-Dimethylnaphthalene		156
15	Methyl-ethylnaphthalene		170
16	Trimethylnaphthalene		170
17	4-Methyldibenzofuran		182
18	1,6-Dimethyl-4-isopropylnaphthalene		198
19	1-Methyl-7-isopropylnaphthalene		184
20	Phenanthrene		178
21	3-Methylphenanthrene		192
22	2-Methylphenanthrene		192
23	9-Methylphenanthrene		192
24	1-Methylphenanthrene		192
25	3,6-Dimethylphenanthrene	(1)	206
	Ethylphenanthrene		206
26	Dimethylphenanthrene	(2)	206
27	Dimethylphenanthrene	(3)	206
28	2,7-Dimethylphenanthrene	(4)	206
29	Dimethylphenanthrene	(5)	206
30	1,6-Dimethylphenanthrene	(6)	206
31	1,7-Dimethylphenanthrene (Pimanthrene)	(7)	206
32	Dimethylphenanthrene	(8)	206
33	Dimethylphenanthrene		206
	Fluoranthene		202
34	Dimethylphenanthrene	(9)	206
35	Pyrene		202
36	Cyclopentanophenanthrene		218
37	Tetrahydroretene		238
38	Retene		234
39	C_3-Phenanthrene		220
40	Chrysene		228

4.2. Influence of Maturity

4.2.1. Changes of retene and tetrahydroretene with increasing depth:

Retene and tetrahydroretene derived from higher plant lipids exist universally in aromatic fractions of the extracts of continental rocks. The PAH of source rocks from two wells(Ai and Sai)are taken as examples. The source rock profiles of the two wells range from immature to mature. The degree

of thermal evolution of their organic matter is not high.
There is not only a rich content of retene but also tetra-
-hydroretene in the immature rocks. The mass spectra of retene
and tetrahydroretene identified for Sai well are shown in Fig.3-1
and 3-2. In order to clearly show the changes in concentration
with increasing depth, we can calculate the concentrations of
individual compounds among all the aromatic compounds from
the GC peak areas. Changes of pattern with depth can be observed
for the two wells in Fig.4-1 and 4-2. From Fig.4-1 and 4-2
it may be seen that the evolutionary characteristics of
retene and tetrahydroretene are similar with increasing depth.

(a) During the immature stage the retene and tetrahydroretene contents of Sai well rise from 5% to 7%, and from 4%
to 6% respectively. The retene and tetrahydroretene contents
of Ai well increase from 6% to 14%, and from 1% to 10%
respectively. It is seen that their contents increase with
increasing depth. Though it is possible that these changes
may be linked with the occurrence of various types of organic
matter, the most likely cause for this early diagenetic
increase in retene and tetrahydroretene is the decarboxylation and aromatization of abietic acid.

(b) During the mature stage the retene and tetrahydroretene contents of Sai well drop from 7% to 1%, and from 4%
to 1% respectively. The retene and tetrahydroretene contents
of Ai well drop from 5% to 1.5%, and from 7% to 1.5%
respectively. Thus it is shown that relative contents decrease
with increasing depth. Apparently, after the maturity threshold these two aromatic biomarkers undergo thermally induced
carbon-carbon bond cracking to form some alkylphenanthrenes
and alkylnaphthalenes with shorter side chains.

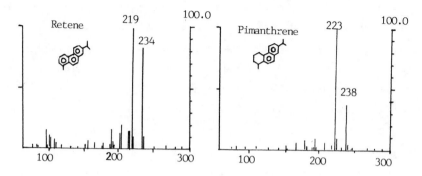

FIGURE 3–1   Mass spectrum of
                    retene

FIGURE 3–2   Mass spectrum of
                    tetrahydroretene

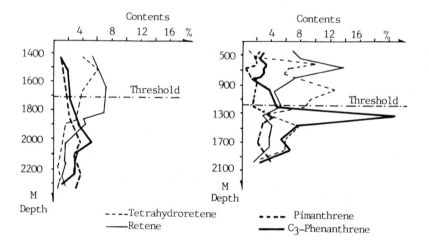

FIGURE 4–1 Depth relationship of
       retene, tetrahydroretene,
       pimanthrene and
       C_3-phenanthrene in Sai well.

FIGURE 4–2 Depth relationship of
       retene, tetrahydroretene,
       pimanthrene and
       C_3-phenanthrene in Ai well.

4.2.2. <u>Changes of Pimanthrene and C_3-phenanthrene</u>
<u>with depth:</u>

Figs. 4-1 and 4-2 illustrate the changes in concentration
of pimanthrene and C_3-phenanthrene with increasing depth.
The GC elution time of the C_3-phenanthrene is very close
to that of retene under our chromatographic conditions
(e.g. Fig. 1), and the changes in the relative concentrations of these two compounds are very significant. While
retene reaches a maximum at rather shallow depths in the
immature section of Sai and Ai wells, Fig. 4-1 and 4-2
respectively, C_3-phenanthrene increases in the immature
section to reach a maximum after the maturity threshold.

The pimanthrene contents of immature rocks in Ai and
Sai wells are all lower than the retene contents and this
is typical of aromatics extracted from continental facies
source rocks. A high abundance of pimanthrene can occasionally appear in humic rocks, but it decreases rapidly
with increasing depth. This means that the rearrangement
of the methyl-substituted position is **thermally**
induced. However, in the Ai and Sai wells the pimanthrene
content increases slightly with increasing depth. Its concentration ranges from about 2% at the immature stage to
3-4% at the mature stage. These concentration changes
indicate that pimanthrene may have two origins. A part can be
derived directly from pimaric acid and another part may be
formed by cracking of some alkylphenanthrene chains.

As mentioned above, in the immature stage, the retene
content increases initially with depth and the pimanthrene
content is more constant. This means that the biological
organic materials are easily transformed into the abietene
carbon skeleton during diagenesis. In the mature stage,

the retene and tetrahydroretene concentrations decrease
with depth but the pimanthrene and C_3-phenanthrene increase.
The data suggest that retene and tetrahydroretene can be
transformed into some smaller alkylphenanthrene molecules
with shorter side chains. Therefore, the changes in these
aromatics with depth can be used as an indicator of thermal
evolution processes and the maturity of source rocks.

4.3. Classification of Organic Matter Types

The existence and abundance of some aromatics often reflect
the precursor organic matter and the environment of deposition. Aromatic extracts of humic type rocks are rich in
phenanthrene, pimanthrene, retene, tetrahydroretene and
chrysene. Sapropel rocks are rich in both alkylphenanthrenes and alkylnaphthalenes. We are mainly concerned
with the changes in content of retene and pimanthrene in the
total alkylphenanthrenes (phenanthrene, methylphenanthrenes
and dimethylphenanthrenes), which can be used as an indicator to recognize different types of organic matter. In
order to study quantitatively the changes of retene and pimanthrene in the total alkylphenanthrenes, the following
four parameters were chosen and twenty-seven aromatic samples have been analyzed. The results are listed in Table 4.

$$\text{Pi}/\bar{\text{p}}(6) = \frac{\text{Pimanthrene}}{\text{Dimethylphenanthrene}(6)} \quad (1)$$

$$\text{Pi}/\text{p}(5) = \frac{\text{Pimanthrene}}{\text{Dimethylphenanthrene}(5)} \quad (2)$$

$$\text{Pi}/\Sigma\text{p} = \frac{\text{Pimanthrene}}{\text{Total Alkylphenanthrenes}} \quad (3)$$

TABLE 4. Four parameter values of retene and pimanthrene in the aromatic fractions of 27 source rocks

Source rocks	Sample No.	Organic type index S_2/S_3	Geological era	Depth(m)	Four parameters			
					Pi/P(6)	Pi/P(5)	Pi/ΣP	Re/ΣP
Immature	Ai-1		Late Cretaceous	450	2.77	1.47	0.16	0.48
	Ai-2			520	3.00	1.33	0.14	0.46
	Ai-3			614	2.50	1.00	0.12	0.64
	Ai-8	$>$ 5.0		970	2.78	1.15	0.12	0.96
	Sai-3			1406	1.58	0.95	0.08	0.50
	Sai-4			1501	1.80	0.93	0.09	0.39
	Sai-5			1660	2.33	1.16	0.10	0.48
Humics Mature	Ai-6		Late Cretaceous	1450	1.94	0.97	0.12	0.34
	Zu-2	$>$ 5.0		1280	1.18	0.87	0.07	0.33
	Du-1		Triassic	2240	3.25	1.86	0.17	0.30
	Du-2		Early Tertiare	2400	5.20	2.36	0.26	0.23
	Du-3			2582	5.10	3.92	0.35	0.15
Mixed	Ai-12		Late Cretaceous	1370	1.16	0.57	0.09	0.23
	Ai-16			1765	1.72	0.95	0.13	0.29
	Sai-7			1848	1.67	0.96	0.12	0.16
	Sai-8	5.0-1.5		1858	1.44	0.93	0.10	0.20
	Sai-9			1875	1.80	1.10	0.13	0.27
	Zu-3		Triassic	1330	1.12	0.73	0.07	0.26
	Na-7		Late Cretaceous	2613	1.40	1.17	0.09	0.27
	Na-8			2660	1.32	0.81	0.13	0.18

TABLE 4.

Sapro-pelics			Age						
Immature	Na-1	< 1.5	Late Cretaceous	1501	0.90	0.55	0.03	0.03
	Na-2			1951	0.88	0.38	0.03	0.02
	Na-3			2171	1.14	0.65	0.06	0.04
Mature	Na-4	< 1.5	Late Cretaceous	2230	1.00	0.55	0.06	0.02
	Na-5			2260	0.97	0.56	0.05	0.02
	Na-6			2351	1.32	0.68	0.10	0.09
	Zu-1		Triassic	1175	0.96	0.65	0.09	0.07

$$Re/\Sigma P = \frac{Retene}{Total\ Alkylphenanthrenes} \quad (4)$$

Σ P=Phenanthrene+Methylphenanthrenes +
Dimethylphenanthrenes

In equations 1 and 2, pimanthrene, dimethylphenanth-
rene(6) and dimethylphenanthrene(5) are analyzed as three
GC peaks eluting close to each other. The different concentrations of the three dimethylphenanthrenes can be directly
shown by GC analysis. The three compounds are all β, β-
dimethylphenanthrenes as described by Radke et al.[3] and
they should have a similar thermal stability. As pimanthrene
is probably derived from diterpenes in conifer resin, the
ratio of pimanthrene to other dimethylphenanthrenes can be
used to recognize conifer input to source rocks. The data of
Table 4 indicate that the values of equations 1 and 2 are
different for various organic facies, with the values of
humic being higher than mixed or sapropelic.

The two values of equations 3 and 4 must be combined
to recognize the organic matter type since some continental
rocks have a high abundance of retene but others have high
pimanthrene contents. Usually, the retene is so rich that
equation 4($Re/\Sigma P$) is more representative. The differences in
the values of equation 4 are very obvious for the three kinds
of organic facies.

In general, the $Re/\Sigma P$ of bitumen derived from sapropelic kerogen is less than 0.10, $Re/\Sigma P$ of the mixed type is
0.1-0.33, and $Re/\Sigma P$ for bitumen from humic kerogen is more
than 0.33. (see Fig.5).

In Fig.5, it is seen that the pimanthrene and retene
parameters for the different types of source rocks have

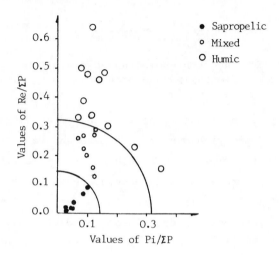

FIGURE 5. Regional distribution map of different source rock
          types.

distinct distributions. The humic range in Fig.5 is bigger
and the sapropelic range is smaller. It seems that either
the Re/ΣP or the Pi/ΣP value is high for humic rocks, but
Re/ΣP and Pi/ΣP are both low for the sapropelic rocks. The
humic range can be separated into two or three parts for the
detailed classification of organic matter type. The values
of these parameters for mature humic rocks are still higher
than those of mixed type rocks and much higher than those of
immature sapropelic rocks. These data indicate that biogenic
marker aromatics can contribute to the classification of
source rocks by kerogen types.

5. CONCLUSIONS

    1. Retene, pimanthrene and the intermediate tetrahydro-
retene originating from higher plant resins, exist universally along with other short chain alkylphenanthrenes in

continental source rocks buried at different depths. In immature humic rocks retene and tetrahydroretene are very abundant, increasing initially with depth, but the content of pimanthrene is stable. In mature source rocks, with increasing depth, the content of pimanthrene increases gradually but the content of retene and tetrahydroretene decreases.

2. As the maturity increases, retene and tetrahydroretene are converted into alkyl phenanthrenes with smaller side chains which increase in quantity with increasing depth.

3. The ratios of retene and pimanthrene respectively to total alkylphenanthrenes and the ratios of pimanthrene to dimethylphenanthrene can be used for the classification of organic matter type.

Acknowledgements

The author sincerely thanks Chief Engineer Wanzhen Lu and Chief Geologist Difan Huang for their advice and Geologists Keming Cheng and Xinzhang Gu for their supply of samples.

REFERENCES

1. D. HUANG, Geochemistry (China), 3, 63-71(1984)

2. M. RADKE, D.H. WELTE, H. WILLSCH, Geochim. Cosmochim. Acta, 46, 1-10 (1982)

3. M. RADKE, H. WILLSCH, D. LEYTHAEUSER and M. TEICHMÜLLER, Geochim. Cosmochim. Acta, 46, 1831-1848 (1982)

4. Z.D. CHENG, The 2nd Organic Geochemical Meeting Beijing, China, 1985(in press)

5. S.G. WAKEHAM, C. SCHAFFNER and W. GIGER, Geochim. Cosmochim. Acta, 44, 403-413 (1980)

6. S.G. WAKEHAM, C. SCHAFFNER and W. GIGER, Geochim.
   Cosmochim. Acta, 44, 415-429 (1980)

7. M.L. LEE, D.L. VASSILAROS, C.M. WHITE and M. NOVOTNY,
   Anal. Chem., 51, 768-774 (1979)

SEPARATION OF VANADIUM AND NICKEL COMPLEXES FROM THE ATHABASCA OIL SAND ASPHALTENES

DESPINA TOOULAKOU and ROYSTON H. FILBY
Department of Chemistry and Nuclear Radiation Center
Washington State University, Pullman, WA 99164-1300

Athabasca oil-sand bitumen was separated into maltenes and asphaltenes by n-pentane precipitation. Greater than 70% of the V and Ni present in the bitumen is found in the asphaltenes as nominally non-porphyrin species. Sequential methanol-acetone extractions of the asphaltenes resulted in reducing the V and Ni associated with the residual solid to 52.8% and 77.8%, respectively. UV-visible absorption spectroscopy and INAA showed that the extracted V was present as vanadyl porphyrins but that non-porphyrin Ni complexes made up most of the extracted Ni component. The data indicate that at least 47% of the V in the Athabasca bitumen is present as vanadyl porphyrin. The association of vanadyl porphyrins with asphaltenes may be an adsorption process, and the solvent extractability of the porphyrins appears to be controlled by the solubility of the asphaltenes in the methanol-acetone solvent.

1. INTRODUCTION

The occurrence of trace metals in petroleum and oil-sand bitumens has been known for a long time. Vanadium and nickel are almost always the most abundant metals (concentrations ranging from 0-3000 ppm), and they play a significant role in petroleum geochemistry. The first important contribution of trace metals to petroleum geochemistry was the identification of metalloporphyrins (vanadyl and later nickel (II) porphyrins) by Treibs[1]. Since that time, metalloporphyrin distributions have become important geochemical parameters in maturation and oil-oil correlation studies[2,3]. Trace element abundances have also been combined successfully with multivariate statistical techniques in oil-oil corre-lation studies[4]. Trace metals in crude oils also cause unfavorable processing effects such as catalyst deactivation and corrosion. Knowledge of the origin and nature of trace metal species in petroleum deposits is thus essential to exploration geochemistry and to the processing of feedstocks.

Vanadium and nickel are present in petroleum almost entirely as organic complexes, but occur only partly as discrete, extractable metalloporphyrins. For example, only 20-25% of the V in the Athabasca oil-sand bitumen can be accounted for in porphyrin complexes; the remainder tradi-tionally is referred to as non-porphyrin species[5] of unde-termined structure. The non-porphyrin Ni and V component is associated with the asphaltenes and represents a highly variable fraction of the total metal content of crude oils. A major reason why the non-porphyrin component of the two most abundant metals remains unidentified is the complexity of the asphaltene structure with which the metal species are associated. Previous studies[6-8] have not been successful in separating metal complexes of unambiguous structures, other than metalloporphyrins, from the asphaltene matrix. Metal species cannot be completely separated chromatographically from asphaltenes because the chromatographic behavior of each fraction is determined by asphaltene functionality and not by the properties of the metal complexes in these frac-tions. In order to determine the structure of the metal species in the asphaltene structure, it is necessary to

separate them from the asphaltenes for identification. The
separation of such metal species from the asphaltene structure will depend on the type and strength of their association with the asphaltenes.

There are a number of possible modes of association for
V and Ni species with the asphaltene structure. The metal
ions (Ni^{2+} and VO^{2+}) may be bonded to functional groups containing S, O, and N in the asphaltene structure, but the
difficulty of asphaltene demetallation suggests that this is
probably not the prominent mode of association, unless the
ligands around the metal ions are in square planar or octahedral geometry. Simple metal complexes, e.g., metalloporphyrins, or other stable complexes, may associate with the
asphaltene structure through π-π and hydrogen bonding. The
strength of these complex-asphaltene associations will
determine whether the metal complexes may be separated from
the asphaltenes by solvent selective extraction or chromatographic techniques. Alternatively, planar metal complexes
may be bound to asphaltenes via a heteroatom-metal bond in
an axial position to give an octahedral complex. Lastly,
there have been reports suggesting that metalloporphyrins,
which are present during the formation of kerogen in the
source rock environment, could be incorporated structurally
into kerogen[2], and subsequently into the asphaltene matrix
during catagenesis, via carbon-carbon linkages to the
metalloporphyrin side chains. Recent spectroscopic studies
provide evidence for either a porphyrin, or at least a 4N,
ligand environment for V in Boscan asphaltenes[9,10,11].
However, it is not possible to identify such complexes or
their mode of association.

In the present study, vanadium and nickel complexes
were separated from the Athabasca asphaltenes by a series

of solvent extraction procedures using methanol-acetone in which asphaltenes are only sparingly soluble. The major advantage of this approach is that extracts containing metal complexes are obtained in an unassociated form and can be identified spectroscopically. Also, the sequential separation of the metal complexes provides information about their mode of association with the asphaltene structure.

2. EXPERIMENTAL

The Athabasca oil sand, from the McMurray Formation, was obtained from the Alberta Research Council. All reagents used were Baker Reagent Grade.

2.1 Separation of Bitumen Components

The separation-extraction procedure used in this work is shown in Figure 1. The bitumen (60 g) was Soxhlet extracted with toluene from batches of Athabasca oil sand (total oil sand extracted was 800 g) in a glass Soxhlet extraction apparatus using cellulose thimbles. The extraction time varied from 6 to 12 hr, depending on the size of the oil-sand sample, and was considered complete when the solvent return was clear. A small amount of finely divided mineral matter (mostly clays) that passed through the cellulose thimbles was separated from the bitumen-toluene solution by centrifugation.

The bitumen was separated into maltenes and asphaltenes by precipitation with a 40:1 volume ratio of n-pentane to bitumen. The precipitation was carried out at room temperature and the solution was stirred for 2 hr. The asphaltenes were separated from the maltene solution by centrifuging at 2500 rpm for 40 min and then vacuum dried (27 mm Hg, 40°C) before reweighing.

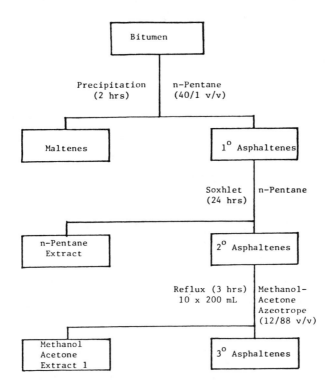

FIGURE 1   Separation Scheme Used for Athabasca Oil Sand
            Bitumen.

An aliquot of the primary asphaltenes was further
Soxhlet extracted with n-pentane for 24 hr to remove residual
resins. A sample of the secondary asphaltenes(7 g) was then
extracted sequentially with ten 200 mL methanol–acetone
azeotrope (12/88 v/v) batches for 3 hr each. At the end of
each stage, the sample was centrifuged at 2500 rpm for
20 min. The supernatent solution was decanted carefully and
the remaining asphaltenes were returned to the extraction
flask with fresh solvent for the next stage. Samples of

solid asphaltenes were taken from each step for Ni and V
determination.   The ten extracts were collected and analyzed
separately and are represented collectively in Figure 1 as
methanol-acetone extract 1.

2.2   Preparation of Extracts

The methanol-acetone extracts were filtered, evaporated in a
Rotovap (27 mm Hg, 35°C) to remove the solvent and were
transferred to preweighed vials where they were allowed to
dry at room temperature.   The samples were then redissolved
in analytical grade tetrahydrofuran (THF) and were made up
to 10 mL in volumetric flasks.   Aliquots of 0.2 mL were
transferred into clean quartz vials where they were diluted
to 5 mL in THF for UV-visible spectrophotometry.   The remaining samples were stored in the dark.   Samples of known
weights of maltenes and n-pentane extract were also made up
to 10 mL in volumetric flasks and were diluted by the same
factor as the other extracts.

2.3   Determination of Nickel and Vanadium

Vanadium and nickel were determined in the bitumen and its
fractions by instrumental neutron activation analysis (INAA)
using methods similar to those described by Jacobs and
Filby[12].

2.4   UV-Visible Spectrophotometric Determination of Metallo-
   porphyrins

The UV-visible spectra of asphaltene extracts and fractions
were obtained with a Perkin Elmer 320 recording spectrophotometer, in the wavelength range of 350-650 nm.

The integrated absorbances (area of metalloporphyrin
peaks) was measured using a method similar to that described
by Sugihara and Bean[13].   The background absorbance contribu-

ting to the Soret band at 408 nm and to the 572 nm band
were subtracted by drawing a base line tangentially through
the valleys on either side. It was necessary to dilute the
solutions ten-fold in order to obtain measurable Soret
absorbances.

3.   RESULTS

The distribution of Ni and V in the components of the Athabasca bitumen are shown in Table I. The weight distribution
of the fractions and the elemental mass balances for Ni and
V relative to the original bitumen are also shown. In
Table II, the Ni and V distributions in the primary asphaltenes (crude) are shown and the mass balances are expressed
relative to the asphaltenes. The Ni and V concentrations in
the total methanol-acetone extract shown in Tables I and II
were obtained by weighted summation of the Ni and V contents
of the ten methanol-acetone extracts of the asphaltenes.
The Ni and V concentrations in the methanol-acetone solutions, solid extracts, and associated asphaltenes are shown
in Table III.

The V content of the maltenes (58.4 μg/g) is equivalent
to 45.3 μgV/g in the bitumen and previous work has shown
that it is present as vanadyl porphyrins associated with
resins[14]. The Ni content of the maltenes, equivalent to
17 μgNi/g bitumen, is much larger than reported nickel porphyrin contents of Athabasca bitumen (2.25-4.94 μgNi/g
bitumen)[15] suggesting that other Ni^{2+} complexes or Ni^{2+}
salts of naphthenic acids as suggested by Jacobs[14] may be
present. Figure 2b shows the UV-visible absorption spectrum
of the maltenes; the characteristic vanadyl porphyrin peaks
at 408, 532, and 572 nm can be clearly identified. The very
small absorbance at 550 nm corresponds to nickel porphyrin

TABLE I   Concentrations (μg/g) and Mass Balances for Nickel and Vanadium in Athabasca Bitumen Components.

Bitumen Component	Mass Balance	Concentration $\overline{X}\pm$s.d.[a] (μg/g)		V/Ni
		Vanadium	Nickel	
Bitumen	100	196±2.54 (100)[b]	64.8±6.58 (100)[b]	3.0
Maltenes	77.5	58.4±0.71 (23.1)	22.0±1.65 (26.3)	2.7
Pentane-Extract	2.77	269±5.56 (3.80)	104±5.62 (4.4)	2.6
Methanol-Acetone Extract	5.1	779±29 (20.3)	177±15 (13.9)	4.4
3° Asphaltenes	15.0	685±22 (52.8)	336±15 (77.8)	2.0

[a]$\overline{X}\pm$s.d. Mean and Standard deviation computed from replicate analyses.

[b]Elemental mass balances in parentheses.

TABLE II   Concentrations (μg/g) and Mass Balances for Vanadium and Nickel in Asphaltene Components.

Asphaltenes Component	Mass Balance	Concentration $\overline{X}\pm$s.d.[a] (μg/g)		V/Ni
		Vanadium	Nickel	
1° Asphaltenes	100	710±21.1 (100)[b]	312±14.2 (100)[b]	2.3
Pentane-Extract	12.4	269±5.56 (4.7)	104±5.62 (4.1)	2.6
Methanol-Acetone Extract	22.8	779±29 (25.0)	177±15 (12.9)	4.4
3° Asphaltenes	66.9	685±22 (64.5)	336±15 (72.0)	2.0

[a]$\overline{X}\pm$s.d. Mean and Standard deviation computed from replicate analyses.

[b]Elemental mass balances in parentheses.

TABLE III Concentrations of V and Ni in Methanol–Acetone Extracts, Solid Extracted Material and Extracted Asphaltenes.

Fraction[a] (wt%)	Concentration in MeOH-Ac Extract		Concentration in Solid Extract[b]		Concentration in Extracted Asphaltenes[c]		V/Ni Weight Ratio	
	V(μM)	Ni(μM)	V(μg/g)	Ni(μg/g)	V(μg/g)	Ni(μg/g)	Extract	Asphaltenes
1 (12.0)	33.9	5.50	850 ± 15	138 ± 4.3	789 ± 29	321 ± 12.5	6.2	2.5
2 (3.58)	10.7	2.69	893 ± 3.1	225 ± 2.9	708 ± 18	281 ± 3.86	4.0	2.5
3 (2.03)	4.90	1.56	726 ± 11	231 ± 14	742 ± 45	319 ± 3.86	3.1	2.3
4 (1.41)	2.96	1.07	628 ± 2.0	238 ± 4.9	736 ± 14	340 ± 3.43	2.6	2.2
5 (0.95)	1.78	0.62	565 ± 28	196 ± 6.8	728 ± 18	314 ± 24.5	2.9	2.3
6 (0.69)	1.23	0.47	538 ± 27	203 ± 7.2	710 ± 15	336 ± 39.8	2.7	2.1
7 (0.58)	1.15	0.45	526 ± 26	207 ± 6.6	795 ± 17	337 ± 7.01	2.5	2.4
8 (0.54)	1.05	0.45	545 ± 29	233 ± 11	762 ± 4.7	342 ± 9.67	2.3	2.2
9 (0.50)	0.95	0.38	527 ± 28	212 ± 7.3	676 ± 8.6	325 ± 13.6	2.5	2.1
10 (0.50)	0.80	0.26	476 ± 26	153 ± 12	685 ± 22	336 ± 15.4	3.1	2.0

[a] Weight percent of initial 2° asphaltenes in extract in parentheses.

[b] Concentration (μg/g) ± standard deviation (counting statistics) except for extracts 1–4 which were calculated from duplicate analyses.

[c] Mean concentrations (μg/g) ± standard deviation calculated from duplicates.

species and indicates that, although the Ni/V ratio is 0.37,
the nickel porphyrin content is small relative to that of
vanadyl porphyrins (assuming similar molar absorptivities),
thus confirming that most of the Ni in the maltenes is present in non-porphyrin Ni^{2+} complexes, e.g., Ni^{2+} naphthenates.

The extraction of the primary asphaltenes with n-pentane
separated 4.7% and 4.1% of the V and Ni, respectively, of
the metal content associated with the asphaltenes. The material extracted by n-pentane (12.4% of the asphaltenes) was
probably coprecipitated resins containing metal complexes.
The UV-visible spectrum of this extract (Figure 2a) is similar to that of the maltenes and vanadyl porphyrin peaks are
the only distinct features, except for the small peak at
550 nm corresponding to nickel porphyrins.

The ten sequential extractions with methanol-acetone
extracted a total of 22.8% of the asphaltene weight. This
amount is not unexpected because asphaltenes contain aromatic
and polar species which may have slight solubility in polar
solvents such as methanol-acetone. The extract contains a
higher V concentration (779 μg/g) than the original asphaltenes, but the Ni concentration is lower. Combined, the
methanol-acetone extracts represent 5.1% of the bitumen by
weight, but 20.3% and 13.9% of the total V and Ni present
in the bitumen, respectively.

The extractability of V and Ni by methanol-acetone from
the asphaltenes are clearly different. The amount of Ni
that remains associated with the residual solid asphaltene
is 77.8% whereas only 52.8% of the V is found associated
with the residual asphaltenes after extraction. This is
difficult to explain on the basis of the selective extraction of vanadyl and nickel porphyrins because the latter

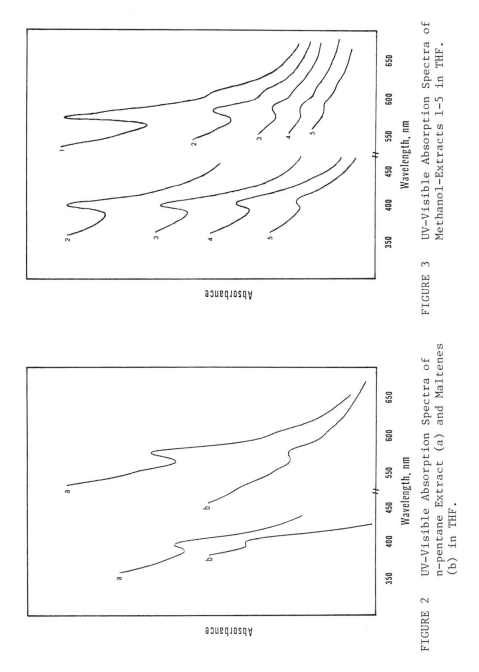

FIGURE 2 UV-Visible Absorption Spectra of n-pentane Extract (a) and Maltenes (b) in THF.

FIGURE 3 UV-Visible Absorption Spectra of Methanol-Extracts 1-5 in THF.

are generally less polar than the corresponding vanadyl porphyrins. A different association of Ni than for V in the asphaltenes is suggested in which the Ni complex exhibits a stronger affinity for the asphaltene matrix than does vanadyl porphyrin; the concept of non-porphyrin Ni complexes is consistent with the small nickel porphyrin content of the maltenes and pentane extract.

The UV-visible absorption spectra of THF solutions of the first five methanol-acetone extracts of the asphaltenes are shown in Figure 3. All spectra reveal the vanadyl porphyrin absorbances at 408 nm, 532 nm, and 572 nm. In contrast, nickel porphyrin is observed only as a very small shoulder at 550 nm in extracts 1 and 2. The absence of nickel porphyrin peaks could be attributed to the high spec-tral background in the region of 550 nm caused primarily by UV-absorbing species in the extracts. However, the Ni content of the methanol-acetone extracts ranges from 16-44% of the V content; thus the very small nickel porphyrin absorbance indicates that non-porphyrin complexes are the major Ni^{2+} species in the extracts. A small peak at 594 nm corresponding to vanadyl rhodoporphyrin is also observed in all the extracts, including the maltenes.

Despite the high absorbance in the 350-670 nm region exhibited by the spectra in Figure 3, the only specific identifiable peaks in this wavelength region are from vanadyl and nickel porphyrins. The lack of other identi-fiable metal species indicates that vanadium is present primarily as vanadyl porphyrins. The absorption spectra of the first five methanol-acetone extracts show a systematic decrease in vanadyl porphyrin absorbances corresponding to the decrease in total V concentration (Table III) and an increase in the background absorbance. The trend continues

through extract 10 until only the Soret peak can be observed.
The UV-visible spectra of the tenth methanol-acetone extract
is shown in Figure 4b, together with the absorption spectrum
of the secondary asphaltenes used in these extractions
(Figure 4a). Calculation of the vanadyl porphyrin content
of the asphaltenes from the small Soret peak shows that it
corresponds to less than 1% of the total V content. The
absorption spectra of the methanol-acetone extracts from
this secondary asphaltene, however, show that substantial
quantities of vanadyl porphyrin can be extracted from the
asphaltene matrix. Thus, spectrophotometric determination
of vanadyl porphyrin in the asphaltene solution severely
underestimates the true vanadyl porphyrin content, as has
been noted previously. Goulon et al.[11] have suggested that
Boscan asphaltene solutions are microheterogeneous and
present supporting ESR evidence. The authors suggested,
that in asphaltene solutions, the molar absorptivity of
metalloporphyrins is a function of the size of the dispersed
asphaltene particles with which the porphyrins are associated. The overall effect is a large reduction of the
metalloporphyrin molar absorptivity relative to the molar
absorptivity of the free complexes.

An attempt was made to determine the vanadyl porphyrin
concentrations of the methanol-acetone extracts. Figures 5
and 6 show the relationship between molar concentration of
V (determined by INAA) and the integrated absorbance of the
408 nm (Soret) and 572 nm peaks, respectively, for the first
five methanol-acetone extracts. Soret absorbances of the
maltenes and n-pentane extract solutions were also measured
and these data are also included in Figure 6. The excellent
correlation of vanadyl porphyrin absorbance with V content
of the extract, including the maltenes and n-pentane extract,

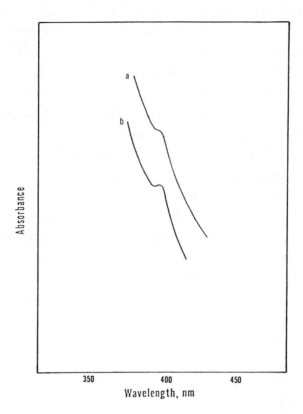

FIGURE 4   UV-Visible Absorption Spectrum of Secondary
            Asphaltenes(a) and Methanol-Acetone Extract
            No. 10 (b) in THF.

suggests that the V in these extracts occurs as metalloporphyrin. Non-porphyrin forms of V are absent, or of low
abundance, unless the ratios of non-porphyrin/porphyrin
species were constant in all extracts, thus giving a linear
Beer's Law plot. This seems unlikely, however, considering
the variation of V content in the extracts and in the
maltenes.

Figure 7 plots the decrease in V content of the
methanol-acetone extracts as a function of the number of

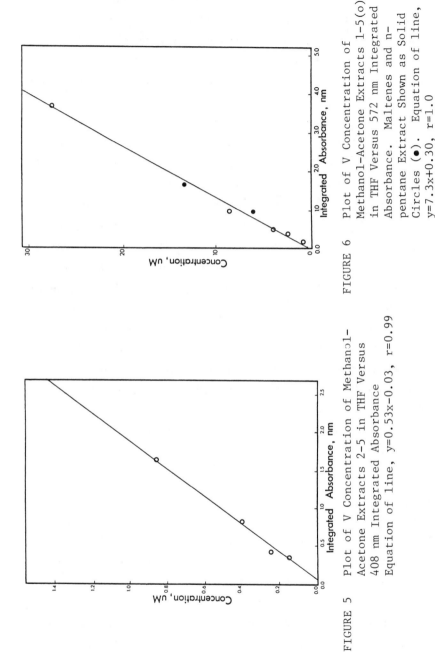

FIGURE 5   Plot of V Concentration of Methanol–
Acetone Extracts 2–5 in THF Versus
408 nm Integrated Absorbance
Equation of line, y=0.53x–0.03, r=0.99

FIGURE 6   Plot of V Concentration of
Methanol–Acetone Extracts 1–5(o)
in THF Versus 572 nm Integrated
Absorbance. Maltenes and n-
pentane Extract Shown as Solid
Circles (●). Equation of line,
y=7.3x+0.30, r=1.0

extractions. The V (and Ni, not shown) concentrations
decrease rapidly in the first four extracts, but then
appear to approach a steady-state or equilibrium value.
Table III also shows that the last five methanol-acetone
extracts approach a "steady-state" solubility of the asphaltenes in this solvent system.

Figure 8 shows the variation in the V/Ni ratio in the
same ten extracts and also the variation in the V/Ni ratio
of the corresponding extracted asphaltenes. The V/Ni ratio
is fairly constant in the solid asphaltenes, with mean
value of 2.2 ± 0.1. The ratio in the extracts, however,
decreases from 6.1 in the first extract to an average of
2.6 ± 0.3 in the last five extracts. These data indicate
that V and Ni species are initially readily extractable and
that VO^{2+} porphyrin is extracted preferentially over the Ni
complex. The fact that the V/Ni ratio in the last five
extracts approaches the value found in the solid asphaltenes
indicates that after removal of readily extracted Ni and V
complexes, the partition of the metal species between the
solid and the solvent may be determined by the solubility of
the asphaltenes in the extracting solvent or the accessibility of the complexes in the asphaltene structure to the
solvent.

4. DISCUSSION

The V and Ni associated with the asphaltenes in crude oils
is conventionally designated non-porphyrin to distinguish it
from metalloporphyrin complexes in the oil. In the Athabasca oil-sand bitumen, the V as vanadyl porphyrin present
in the maltenes and n-pentane extract of the asphaltenes
represents 26.9% of the V in the bitumen. The corresponding
extractable Ni content is 30.7% but only a small fraction

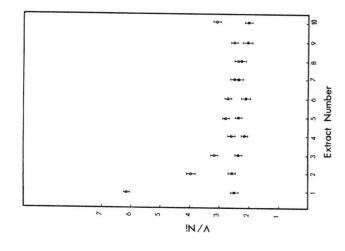

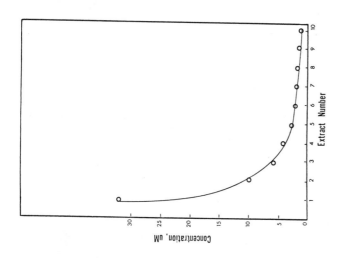

FIGURE 7 Variation of V Concentration (μM) in Methanol-Acetone Extract Solutions with Extract Number.

FIGURE 8 Variation of V/Ni Ratio in Methanol-Acetone Extracts (open circles) and Corresponding Asphaltenes (solid circles).

appears to be present as nickel porphyrin with the remaining amount present as other Ni^{2+} complexes, possibly as Ni^{2+} naphthenates.

In the case of V, it is shown that at least 25% of the nominally non-porphyrin V in the asphaltenes is present as vanadyl porphyrin. After the extraction of this vanadyl porphyrin component, the asphaltenes still contain vanadyl porphyrins as shown spectroscopically. Hence, for the Athabasca bitumen, at least 92.5 μgV/g bitumen (or 47% of the total V) is present as vanadyl porphyrin, whereas previous determinations of the vanadyl porphyrin content of the Athabasca bitumen range from 38.6-59.5 μgV/g bitumen.[14] It is not clear, however, if all of the V remaining in the asphaltenes is present in porphyrin species or whether non-porphyrin V complexes are present. Several recent spectroscopic studies of Boscan asphaltene[9,10,11], however, suggest that all V is present in a planar 4 N ligand environment-- probably porphyrinic. Similar conclusions regarding Ni cannot be made, however, from these data.

The methanol-acetone extraction data of Athabasca asphaltenes suggest a model for V behavior:

Asphaltene Precipitation: During precipitation of asphaltenes by n-pentanes, vanadyl porphyrins are partitioned between the maltenes and asphaltenes. The association of vanadyl porphyrins with the asphaltenes may be adsorption or bonding to polar molecules in the asphaltenes via a V-N or V-O bond. At the time of precipitation, some vanadyl porphyrin will be co-adsorbed with polar resins in the asphaltenes.

Asphaltene Extraction: Extraction of asphaltenes with n-pentane removes co-precipitated resins and associated vanadyl porphyrin. Vanadyl porphyrin adsorbed or bonded to

the more polar asphaltene molecules is not extracted with
the low polarity solvent. With the more polar methanol-
acetone solvent, however, adsorbed vanadyl porphyrin is
extracted, but as the virtual constancy of the V concentration in the solid extract shows, the concentration of VO^{2+}
porphyrin is controlled by the solubility of the asphaltenes
in the solvent. The partitioning of vanadyl species between
the asphaltenes and the solvent, then, may depend on solvent
accessibility which is determined by asphaltene solubility.
A similar argument may apply to Ni species, although apparently both porphyrin and non-porphyrin species are involved.

REFERENCES

 1. A. TREIBS, Angew. Chemie 49 682 (1936).
 2. E.W. BAKER and J.W. LOUDA, in Advances in Organic
    Geochemistry, 1981, Edited by M. Bjorøy (John Wiley
    and Sons, 1983), pp. 401-421.
 3. A.J.G. BARWISE and I. ROBERTS, Org. Geochem. 6 167
    (1984).
 4. B. HITCHON and R.H. FILBY, AAPG Bull. 68 838 (1984).
 5. F.S. JACOBS, F.W. BACHELOR, and R.H. FILBY, in
    Characterization of Heavy Crude Oils and Petroleum
    Residues (Editions Technip, Paris, 1984), pp. 173-178.
 6. W.A. SPENCER, J.F. GALOBARDES, M.A. CURTIS, and L.B.
    ROGERS, Sep. Sci. Technol. 17 797 (1982).
 7. R.H. FISH, J.J. KOMLENIC, and B.K. WINES, Anal. Chem.
    56 2452 (1984).
 8. F.S. JACOBS and R.H. FILBY, Am. Chem. Soc. Div. Petrol.
    Chem. Preprints 28 758 (1983).
 9. V.M. MALHOTRA and H.A. BUCKMASTER, Fuel 64 335 (1985).
10. C. BERTHE, J.F. MULLER, D. CAGNIANT, J. GRIMBLOT, and
    J.P. BONNELLE, in Characterization of Heavy Crude Oils
    and Petroleum Residues (Editions Technip, Paris, 1984),
    pp. 164-168.
11. J. GOULON, A. RETOURNARD, P. FRIANT, C. GOULON-GINET,
    C. BERTHE, J.F. MULLER, J.L. PONCET, R. GUILARD, J.C.
    ESCALIER, and B. NEFF, J. Chem. Soc. Dalton Trans.
    1095 (1984).
12. F.S. JACOBS and R.H. FILBY, Anal. Chem. 55 74 (1983).

13. J.M. SUGIHARA and R.M. BEAN, J. Chem. Eng. Data 7 269 (1962).
14. F.S. JACOBS, Ph.D. Dissertation, Washington State University, Pullman, WA, U.S.A. (1982).
15. S.K. HAJIBRAHIM, Proc. 13th Int. Symposium on Chromatography, Canne, France (1980).

USING MEAN SQUARE DIFFERENCE OF STERANES AND TRITERPANES TO CORRELATE CRUDE OILS AND SOURCE ROCKS FROM WESTERN-LIAOHE DEPRESSION

NAIHUANG JIANG
Scientific Research Institute of Petroleum
Exploration and Development
P. O. Box 910
Beijing, China

The correlation of crude oils and source rocks using
sterane and triterpane biomarkers is often performed
by visual comparison of chromatograms from gas
chromatography-mass spectrometry. It has been improved by a simple mathematical mean square difference
(m.s.d.) analysis. The m.s.d. is used to determine
the relationship between two curves in a coordinate
in mathematics. As demonstrated here the method adds
precision and accuracy in evaluating relative correlation strengths when several correlation sets are
being compared. It is easy to accomplish and utili-
ze.

1. INTRODUCTION

The purpose of crude oil/source rock geochemical correlation is to identify the oils and their source beds. This correlation proves, suggests or disproves a relationship between
the reservoired oils and source beds. It is very useful for
petroleum exploration and the genetic study of petroleum.

There are many methods used for correlation, but one of
the best correlation tools is provided by geochemical fossils such as steranes and triterpanes. Based on GC-MS analyses of oils and source rocks we have drawn, certain diagrams of sterane and triterpane diagnostic molecular distri-

butions. The similarity between these diagrams of the oils
and source rocks is recognized as a basis of correlation.
GC records of the saturated hydrocarbon fraction and certain
mass chromatograms such as m/z 191 (for triterpanes) and
m/z 217 (for steranes) are often directly used for correlation. However the similarity which is mainly estimated
visually is too rough for more detailed identification. By
using the mean square difference the similarity can be defined in detail.

2. CALCULATION OF M.S.D.

The calculation of mean square difference is simple(Fig.1).
Assume there are n components $S_1, S_2, S_3, \ldots \ldots S_n$. The corresponding abundances are $a_1, a_2, a_3, \ldots \ldots a_n$ for the source rock sample and $a'_1, a'_2, a'_3, \ldots \ldots a'_n$ for the oil sample.
The mean square difference of steranes and triterpanes
between the source rock sample and the oil sample is defined as:

$$A = \sqrt{\frac{\sum_{i=1}^{n} (a_i - a'_i)^2}{n}}$$

Where A is the mean square difference (m.s.d.). Consequently, the smaller the A, the better the correlation.

3. DISCUSSION OF OIL/SOURCE ROCK CORRELATION USING M.S.D.

Mean square difference is used to determine the correlation
of two curves in mathematics. If a curve fully coincides

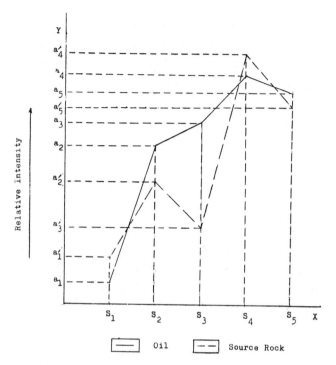

FIGURE 1 Schematic diagram of diagnostic sterane or triterpane molecular distribution.

with another one, the m.s.d. must be 0, the smaller the m.s.d.
the more coincident the correlation.

   We use diagnostic sterane and triterpane molecular ions
from GC–MS analysis as the components of the m.s.d. analysis.
Their corresponding intensities in the oil sample analysis
are used to draw a curve and those from the source rock sample to draw another curve. The exercise is repeated using
the appropriate GC–MS sterane and triterpane diagnostic ions.
Thus, the numerical value of m.s.d. of steranes and triterpanes can be calculated and on the basis of both numerical
values, we can identify the oils and their source beds.

Correlating the degree of oil/source rock correlation is
defined more precisely by m.s.d. than by visual comparison.

D. Leytheauser et al (1977) have done experiments on
three groups of oil and source rock samples. In the first
group the source rock and the reservoir are located in same
bed (Fig. 2I). In the second group, the source rock is
next to the reservoir (Fig. 2II). In the third group, the
source rock is at some distance from the reservoir (Fig. 2 III).
In Fig. 2, it can be seen that the pairs in the first group
are in good correlation, as compared to the second group
and the third group is the worst. This does not mean that
all the pairs within one group have the same degree of the
correlation. We can not differentiate further degrees of
correlation within each group visually. For example,it is
difficult to visually compare the pairs D and B in the first
group, and the pairs A and E in the second group.
If m.s.d. is used, we can easily measure the degree of
correlation. For instance, the numerical value of m.s.d. of
the pairs C, F, E, A, in the second group (Fig. 2II) are 3.04
6.61, 8.81, 9.88, respectively, indicating the best correlation for C and the worst for A.

4. USING M.S.D. IN OIL/SOURCE ROCK CORRELATION

The western Liaohe Depression is one of the main oil producing basins in China. The six samples (three oil samples and
three source rock samples) in this paper are chosen from
Gaosheng and Shuguang oil field in this Depression. Their
sampling location is shown in Fig. 3.

The maturity of the source rocks and crude oils is determined by the 22S/22R ratio of $17a(H)$-homohopane. A ratio
of 1.0 is used as the threshold value.

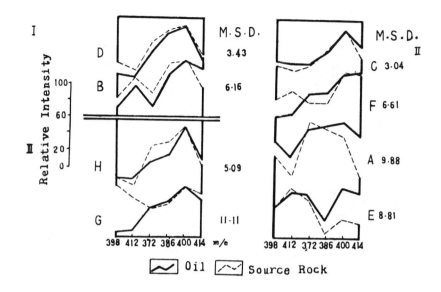

FIGURE 2   Determination of degree of correlation between oils and source rocks based on distribution of C_{27}^{+} cyclic hydrocarbons and corresponding m.s.d..

The Roman members are explained as follows:

    I. the first group of pairs: the source rock and the reservoir of the oil are located in the same bed.

    II. the second group of pairs: the source rock is next to the reservoir of its oil.

    III. the third group of pairs: the source rock is at some distance from the reservoir of the oil.

(The diagram is from " Source rock/crude oil correlation based on distribution of C_{27}^{+} cyclic hydrocarbons " by D. Leythaeuser, A. Hollerbach and H. W. Hagemann (1977) in Advances in Organic Geochemistry, 1975, R. Campos and J. Goni eds., pp 3-20, Enadimsa, Madrid. The numerical m.s.d. values are added by the author.)

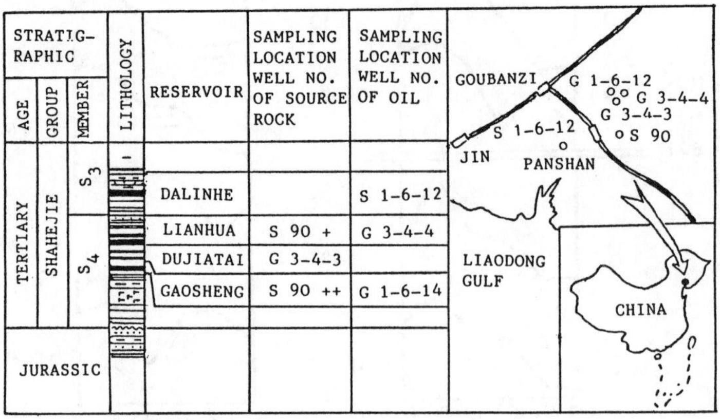

FIGURE 3 The sampling location of oils and source rocks.

+ the sample no. 80117.

++ the sample no. 80118.

Of the three source rock samples, only the sample at 2790m
(sample no. 80118) a black shale from the 4th member of
Shahejie Formation (S_4) of well no. S 90 is mature. The other
two samples, a dark grey shale (sample no. 80117) from S_3 at
2305.5m and another dark grey shale from S_4 at 1875m of well
no. G 3-4-3 are immature.

Of the three oil samples, one of them is from S_4 of
Gaosheng reservoir well no. G 1-6-14, and one is from S_3 of
Lianhua reservoir, well no. G 3-4-4. The above two wells
are located in the Gaosheng oil field, and the remaining one
is from S_4 Dujiatai reservoir and S_3 Dalinhe reservoir of
well no. S 1-6-12 in the Shuguang oil field.

In order to identify the oil and its source bed, the two
diagrams of sterane and triterpane diagnostic molecular dis-

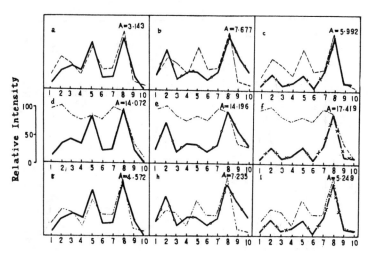

FIGURE 4 Oil/source rock correlation by diagnostic sterane
molecular distribution pattern.

──────────── extract of source rock, S_4 member of well no. S 90.

─o─o─o─ extract of source rock, S_3 member of well no. S 90.

─+─+─+─ extract of source rock, S_4 member of well no.
            G 3-4-3.

─ ─ ─ ─ oil from S_4 member of well no. G 1-6-14.

─·─·─·─ oil from S_3 member of well no. S 1-6-12.

─··─··─ oil from S_3 member of well no. G 3-4-4.

The numbering on the X axis in this diagram is as follows:
1. 5α-cholestane 20S 2. 5α-cholestane 20R 3. 24—methyl –
cholestane 20S 4. 24-methyl-coprostane 20R 5. 24—methyl-
cholestane 20R 6. 24-ethyl-cholestane 20S 7. 24-ethyl –
coprostane 20R 8. 24-ethyl-cholestane 20R 9. 4-methyl –24–
ethyl-cholestane (isomer) 10. 4-methyl-24-ethyl-cholestane.
A = mean square difference.

tributions are shown in Fig. 4 and Fig. 5. Their m.s.d. are
listed in Table 1.

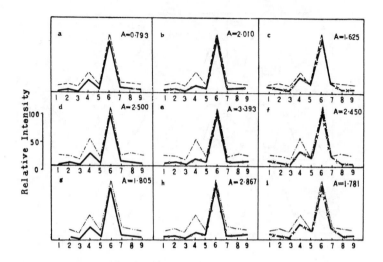

FIGURE 5 Oil/source rock correlation by diagnostic triterpane molecular distribution pattern.

——————— extract of source rock, S_4 member of well no. S 90.

—o—o— extract of source rock, S_3 member of well no. ·S 90.

—+—+— extract of source rock, S_4 member of well no.
         G 3-4-3.

— — — oil from S_4 member of well no. G 1-6-14.

—·—·— oil from S_3 menber of well no. S 1-6-12.

—··—··— oil from S_3 member of well no. G 3-4-4.

The numbering on the X-axis in the diagram is defined as
follows:

1. $18\alpha(H)$-22,29,30-trinorhopane 2. $17\alpha(H)$-22,29,30-trinor-
hopane 3. $17\alpha(H)$-29,30-binorhopane 4. $17\alpha(H)$- 30-norhopane
5. $17\beta(H),21\alpha(H)$- 30-normoretane 6. $17\alpha(H)$-hopane 7. 17β
$(H),21\alpha(H)$-moretane 8. $17\alpha(H),21\beta(H)$-30-homohopane 22S
9. $17\alpha(H),21\beta(H)$-30-homohopane 22R.

A = mean square difference.

In Figs. 4, 5 and Table 1 the correlation degree can be

divided into three classes as follows according to their
m.s.d..

The first class, excellent correlation (m.s.d. of steranes < 4, m.s.d. of triterpane < 1.0):

An excellent correlation is observed between the S_4
member matured source rock sample (no. 80118) and the oil
sample from the S_4 Gaosheng reservoir of well no. G 1-6-14.
Both their sterane and triterpane diagnostic molecular distribution patterns are very similar to each other. The
m.s.d. of both sterane and triterpane are the least among
all the pairs. This may be attributed to the lenticular
distribution of Gaosheng reservoir within the S_4 source
beds. Thus, the pooled oil is uncontaminated by oils
from other probable source beds. The excellent correlation
is coincident with the geological interpretation.

The second class, moderate correlation (m.s.d. of steranes from 4 to 6 and m.s.d. of triterpanes from 1 to 2):

The correlation of the oil from S_4 member Lianhua reservoir of well no. G 3-4-4 with the three source rock
samples is moderate and there are some differences among the
three pairs. The correlation of the oil with the mature
source rock sample (no. 80118) is better than with those
of lower maturity . We suggest that the oil could have
been mainly derived from the mature source beds.

The correlation of the oil sample from S_4 Gaosheng reservoir with two source rock samples from S_4 and S_3 having
low maturity belongs to the second class too. As mentioned
above, this oil is mainly drived from mature source beds,
but a small portion of it may have been derived from the
low mature source beds.

The third class, bad correlation (m.s.d. of steranes

TABLE 1   The m.s.d. of diagnostic sterane and triterpane molecular distributions.

Oil	M.S.D. of Source Rock					
	S$_4$ Mature of S 90		S$_3$ Low Mature of S 90		S$_4$ Low Mature of G 3-4-3	
	(Steranes)	(Terpanes)	(Steranes)	(Terpanes)	(Steranes)	(Terpanes)
S$_4$ Gaosheng +	3.143	0.793	7.677	2.010	5.992	1.625
S$_3$ Dalinhe & Dujiatai ++	14.072	2.500	14.196	3.393	17.419	2.450
S$_3$ Lianhua +++	4.572	1.805	7.235	2.867	5 249	1.781

M.S.D.= mean square difference, see the text.

+ The oil from well no. G 1-6-14.

++ The oil from well no. S 1-6-12.

+++ The oil from well no. G 3-4-4.

> 8, m.s.d. of triterpanes > 2).

The correlation of the oil from both S_3 Dalinhe reservior and S_4 Dujiatai reservoir of well no. G 1-6-12 with all of the three source rock samples belongs to this class. The curves of the three pairs are shown by d, e and f in Figs. 4 and 5. Both the sterane and triterpane distribution curves of the oil deviate greatly from those of the three source rock samples as compared with the other pairs. We suggest that this oil could have been derived from unknown source beds, other than the mature or low mature source beds of S_3 and S_4 in the area near Gaosheng oil field in the central part of this depression.

In summary, we have demonstrated the mean square difference method of correlating sterane and triterpane diagnostic molecular distribution patterns. It is more effective and precise than visual methods in evaluating oil/source rock correlations.

ACKNOWLEDGEMENT

I am grateful to Mr. Wang Baozhi of the Research Institute of Liaohe oil field for sampling.

DETERMINATION OF BIOMARKERS IN GEOLOGICAL SAMPLES BY TANDEM MASS SPECTROMETRY

R. P. Philp
School of Geology and Geophysics
University of Oklahoma
Norman, Oklahoma 73019 USA

M. Johnston
Finnigan MAT
San Jose, California   USA

The use of biomarkers as an aid to petroleum exploration is well documented in the literature. Normally biomarker distributions are determined by gas chromatography-mass spectrometry (GC-MS) using single ion monitoring and multiple ion detection. The GC step generally requires fractionation of the sample prior to analysis. This paper describes an alternative and rapid method to determine the distribution of biomarkers in unfractionated crude oils. The method uses tandem mass spectrometry (MS/MS) and involves the monitoring of parent/daughter ion relationships for various groups of biomarkers. The resulting parent ion distributions permit a rapid screening of oils to determine those derived from similar or different sources. Major advantages of the method are speed of analysis, no requirement for fractionation, and ability to determine distributions of different classes of biomarkers which are not readily separable by GC-MS. A major disadvantage is the inability to resolve stereoisomers. However the rapidity of the analysis means that after the initial screening only a relatively small number of samples need to be analysed by GC-MS or GC-MS/MS. This paper describes examples of the various biomarker distributions in crude oils as determined by MS/MS and GC-MS/MS. It also demonstrates that many classes of biomarkers extend far above the carbon number range previously determined by GC-MS alone.

1. INTRODUCTION

The past ten years have seen a massive explosion in the applications of gas chromatography-mass spectrometry (GC-MS) to the determination of organic compounds, referred to as biomarkers, in geological samples and in particular fossil fuels[1]. The determination of biomarkers has proved extremely valuable in many petroleum exploration problems. Biomarkers are organic compounds, generally with complex structures, that occur in fossil fuels but can be related to specific precursor molecules present in particular source materials[2]. Once a precursor/product relationship is established it permits information to be obtained on the nature of the organic material in a source rock and in turn whether a particular source rock has produced, or is likely to produce, oil or gas. Furthermore, stereochemical changes occur to these molecules as a result of burial and elevated temperature and can be used to determine the maturity of the material. In addition the biomarker distributions can yield information on biodegradation and relative migration distances. Comparison of biomarker fingerprints from different samples permits oil/oil and oil/source rock correlations to be undertaken. These applications have been well documented in a number of review articles and research papers, and some of the more recent ones are cited here[1-5].

The method of choice for these analyses has been gas chromatography-mass spectrometry (GC-MS) with particular emphasis being placed on the use of single ion monitoring (SIM) or multiple ion detection (MID) techniques. However, the requirement for sample fractionation and subsequent gas chromatographic separation is a very time-intensive step, taking several hours for each analysis to be completed. In this paper we present an alternative method for determining the biomarker distributions in crude oils and source rock extracts which permits whole oils and extracts to be analysed in a few minutes rather than hours as with GC-MS. Tandem mass spectrometry or MS/MS and sample introduction by way of the direct insertion probe is the method to be described. MS/MS is a relatively recent development as far as commercial instrumentation is concerned and few applications to biomarker characterization have been previously published in the literature. In this paper, a brief summary of the MS/MS technique as it is used for biomarker analysis will be given, followed by a number of examples where it has been used for the purposes of determining biomarker distributions in unfractionated oils and extracts.

1.1   Tandem Mass Spectrometry (MS/MS)

A variety of MS/MS configurations are currently available. A Triple-Stage Quadrupole (TSQr) was used for this work, which provides unit mass resolution to 1800 amu in both stages of mass analysis.

Triple quadrupole instruments have two quadrupole mass analyzers separated by an encased quadrupole collision cell (Fig. 1). Ions selected by the first quadrupole mass analyzer (termed "parent ions") are transmitted into the collision cell and allowed to undergo collisions with an inert gas such as argon or nitrogen. The "collisionally activated" parent ions can decompose to form new ionic species termed "daughter ions". The quadrupole within the collision cell is not operated as a mass analyzer, but

TRIPLE STAGE QUADRUPOLE GC/MS/MS SYSTEM

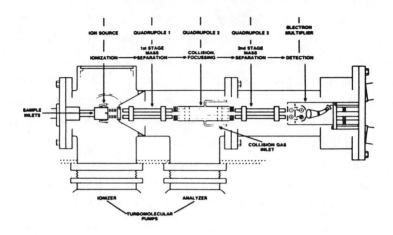

FIGURE 1.   Diagram showing the arrangement of the
        quadrupoles in the triple stage quadrupole system
        used in this study.

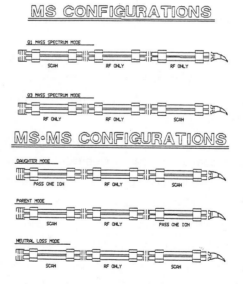

FIGURE 2.   Diagrammatic representation of the different
        operating modes for the MS/MS system.

serves to reduce the scattering of the daughter ions, and allows them to be transmitted to the third quadrupole. Mass analyzed daughter ions are then detected by an electron multiplier.

Several different operation modes allow the analyst to selectively acquire a subset of the large data set that MS/MS systems generate (Fig. 2). By scanning either the first mass analyzer (Q1) or the second mass analyzer (Q3) with no collision gas present in the middle quadrupole region, mass spectra can be obtained which are virtually identical to those generated by single stage quadrupole mass spectrometers.

To obtain the mass spectrum of the fragments of a particular parent ion, daughter scans are acquired. The first mass analyzer (Q1) is set to pass parent ions of a selected mass, the collision cell is pressurized, and the second mass analyzer (Q3) is scanned over a mass range appropriate for the transmission of daughter ions. The resulting spectrum can give valuable structural information and may be used to uniquely identify the parent ion.

To obtain the spectrum of ions resulting from the loss of a neutral moiety in the collision cell, both mass analyzers are scanned in concert. A mass offset is held between the analyzers such that only ions that lose the selected neutral fragment can be transmitted through both analyzers and be detected. Such neutral loss spectra may be used to detect the presence of classes of compounds whose decomposition pathways involve removal of neutral fragments.

If the first mass analyzer is scanned while the second mass analyzer is set to transmit a selected mass, parent spectra are produced. A parent spectrum shows the masses

of all ion source products that fragment to produce a
selected daughter. Parent spectra may be used to identify
homologous compounds that have in common a core
substructure.

1.2 MS/MS of Biomarkers

Several important classes of biomarkers are amenable to
analysis using parent scans. For example, if the molecular
ions of any of a series of hopanes are transmitted into the
collision cell of the triple-stage quadrupole system (e.g.,
a daughter scan of the molecular ion is acquired), a major
fragment formed by the collisionally-activated
decomposition (CAD) is m/z 191 (Fig. 3). Every member of
the hopane series thus far examined has produced an intense
m/z 191 fragment.

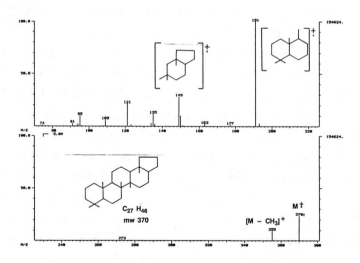

FIGURE 3.   The parent ion of the C_{27}-hopane, m/z 370,
undergoes a CAD to produce an intense daughter ion at
m/z 191.

By employing parent scans of m/z 191, an idea of the content and distribution of hopanes in a sample may be gained rapidly and specifically, without any prior extraction or chromatographic separation. Similar experiments performed on other terpanes and steranes indicate that the distribution of these biomarkers may also be determined by MS/MS.

By carefully controlling the instrumental parameters, a wealth of information may be obtained in a few minutes. For example, the degree of fragmentation in the collision cell may be varied by changing the energy at which parent ions are accelerated into the collision region. Fig. 4a illustrates the effect of 10 eV collisions on the M-1 ion generated in the CI analysis of a C30-hopane. The resulting daughter ions are distributed towards the lower half of the mass spectrum. In contrast, the 30 eV daughter spectrum in Fig. 4b indicates the production in the collision cell of fewer of the higher-mass fragments.

The classes of biomarkers that are most often monitored for correlation purposes include isoprenoids, sesquiterpenoids, diterpenoids, tricyclic terpanes, pentacyclic terpanes and steranes[2]. It is possible, with appropriate selection of parent and daughter ions, to monitor the distribution of virtually all of these classes of biomarkers in one direct insertion probe MS/MS analysis of an unfractionated oil or rock extract. The resulting distributions obtained from these analyses are basically molecular ion distributions. At the present time it is not possible to resolve stereoisomers in classes of biomarkers such as the steranes or triterpanes using this technique. However, it is not felt that this is a critical limitation because the MS/MS method is being utilized as a screening

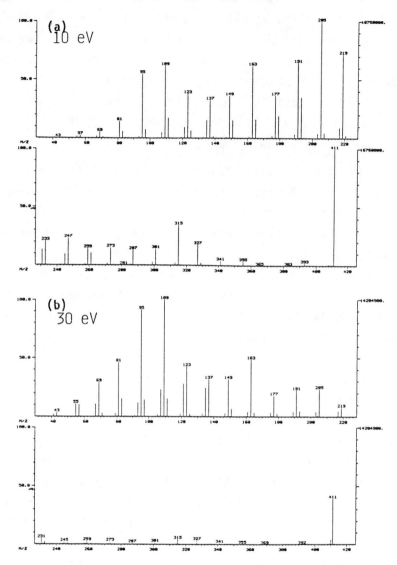

FIGURE 4.   Comparison between the collision activated
decomposition spectrum for the M-1 ion of the C_{30}
hopane with argon obtained at (a) 10 and (b) 30 eV.
The M-1 ion was used in this particular study since
the ion source was being operated in the chemical
ionization mode.

technique to determine oils derived from similar types of source materials which in turn can be expected to have similar molecular ion distributions. Samples can then be selected from this screening process and analysed in more detail by GC-MS or GC-MS/MS. The use of the MS/MS screening technique will greatly reduce the number of samples that must be analysed by GC-MS and thus greatly shorten analysis times.

In the following sections examples are given demonstrating the use of MS/MS in biomarker analyses. A number of different oils have been used to obtain these results. Some samples were run with the ion source operating in the electron impact mode and others in the chemical ionization mode, and this information is given in the figure captions. The unfractionated oils are introduced directly into the ion source of the mass spectrometer by steadily heating the direct insertion probe. Various parent/daughter ion relationships were monitored and are described in more detail below, but in all cases the collision energy was set at 30 eV and argon was used as the collision gas.

2. RESULTS AND DISCUSSION

This paper represents one of the first applications of MS/MS to biomarker analysis since the work of Gallegos[6]. A great deal of experimentation remains to be undertaken. It is hoped that analyses of the CAD spectra for the different classes of biomarkers will reveal differences in the daughter ion spectra that can be used to provide unique parent/daughter ion relationships that will distinguish between classes of biomarkers with spectra that contain

many ions in common such as alkanes and isoprenoids. This
will require the examination of the CAD spectra for pure
standards of the different classes of biomarkers obtained
at different collision energies to determine any possible
significant differences in the daughter ion spectra of
these compounds.

2.1 Biomarker distributions by MS/MS

The first example demonstrates an experiment to measure the
parent ions that produce a daughter ion at m/z 183, which
is a characteristic fragment of both alkanes and
isoprenoids. The resulting distribution of parent ions of
m/z 183 produced by the direct analysis of the
unfractionated oil shows the presence of a homologous
series up to at least C_{50} and undoubtedly extending beyond
this value (Fig. 5a). A comparison with the chromatogram
obtained from the analysis of the same oil using on a flame
ionization detector (Fig. 5b) clearly shows the enhancement
of the n-alkane distributions above C_{30} obtained when using
the MS/MS technique with the direct probe analysis. This
reflects one limitation of GC-MS, namely the inability to
get high molecular weight components through the column
because of problems with liquid phase stability at high
temperatures. A limitation of monitoring parent ions of
the m/z 183 daughter ion is that it does not distinguish n-
alkanes from isoprenoids but   examination of the
chromatogram in Fig. 5b indicates that the contribution
from isoprenoids above C_{20} will be minimal.

2.1.2 Sesquiterpanes

Sesquiterpanes are a class of bicyclic biomarkers that have
been found in many oils derived from terrigenous source
material[7]. These compounds occur as complex mixtures of
C14, C15, and C16 homologs and isomers. Electron
ionization of sesquiterpanes yields intense fragments at
m/z 109 and m/z 123 and collision-activated decomposition
of EI molecular ions also produce abundant daughter ions at
these masses. Figs. 6a and b show the parent ion
distribution responsible for the daughter ions at m/z 123
and m/z 109, respectively.

In this example the problem is made more complex by
the fact that a number of isomers and isobars may
contribute. Hence the peak at m/z 194 corresponds to all
the C_{14} sesquiterpanes which produce a daughter ion at m/z
123 (Fig. 6a) or m/z 109 (Fig. 6b). The fact that the
parents of m/z 109 show a different distribution from m/z
123 is due to different isomers of the same homologue
having different intensities of the daughter ions at m/z
109 and 123. Hence when the C_{14} parents of m/z 109 and m/z
123, i.e. m/z 194, are separately summed, the resulting
total intensity of the parent ion will be different for the
two daughter ions. The fact that the various compounds
make different contributions to the m/z 109 and m/z 123
parent spectra may eventually permit differentiation of the
contributions of the various stereoisomers.

2.1.3 Tri-, tetra-, and pentacylic terpanes

Polycyclic terpanes are ubiquitous in geological samples
and there are many reports in the literature of the use of

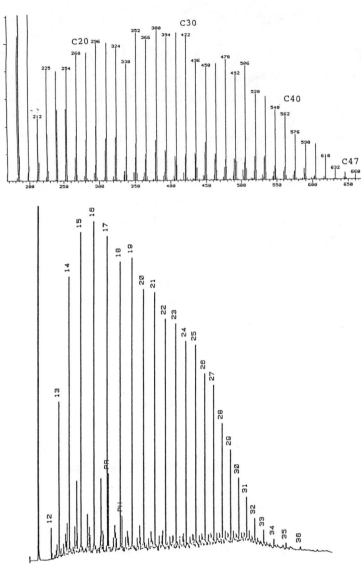

FIGURE 5.   (a) This chromatogram shows the distribution of the parent ions of m/z 183 which are molecular ions for the n-alkanes plus isoprenoids. (b) A comparison with the GC analysis shows how direct insertion probe MS/MS extends the range of components detected in a mixture.   The C_{47} alkane parent ion is the maximum in Fig. 5 whereas it is n-C_{36} in Fig. 5b.

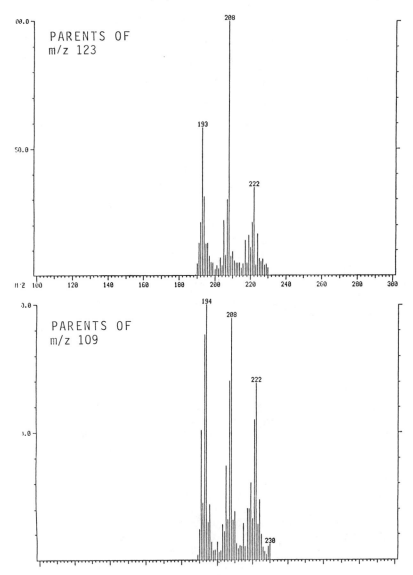

FIGURE 6a and b. The daughter ions at m/z 109 and 123 are
        significant in the CAD spectra of bicyclic
        sesquiterpanes. The distributions of their parent
        ions, shown here, differ since the intensity of the
        daughter ions is different in the various
        stereoisomers of the same homologue.

these compounds in petroleum exploration studies. Molecular ions of the tri-, tetra- and pentacyclic terpanes (whose molecular formulas correspond to $C_nH_{(2n-4)}$, $C_nH_{(2n-6)}$, and $C_nH_{(2n-8)}$, respectively) have all been shown to produce abundant daughter ions at m/z 191 upon collisional activation. Parent scans of m/z 191 can therefore be used to determine the relative abundances of these compounds (Fig. 7). A vast amount of information is present in the distribution of these ions, and complements the information

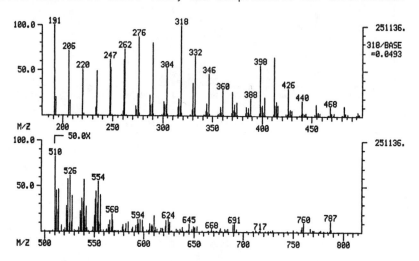

FIGURE 7.   Tricyclic, tetracyclic and pentacyclic terpane CAD spectra all have strong daughter ions at m/z 191. The determination of the parents for this daughter ion typically produces a parent ion distribution of the nature shown in this figure. The advantage of this data over GC-MS data is the fact that the distribution of all three terpane families are easily resolved. The extension of these distributions into higher molecular weight regions can also be observed. The disadvantage is the fact that stereoisomers and diastereomers cannot be resolved. Tricyclic terpanes have molecular ions at C_nH_{2n-4} (i.e. $C_{30}H_{56}$, m/z 416); tetracyclic terpanes C_nH_{2n-6} (i.e. $C_{30}H_{54}$, m/z 414) and pentacyclic terpanes C_nH_{2n-8} (i.e. $C_{30}H_{52}$, m/z 412).

that may be obtained from GC/MS and single-ion monitoring.

In order to determine the tricyclic terpane distribution by GC/MS it has thus far been necessary to isolate a fraction containing the tricyclics, since above C_{27} the hopanes interfere in the chromatogram[8]. Analysis of such a fraction allows observation of the tricyclic terpanes up to the C_{45} homolog with chromatographic

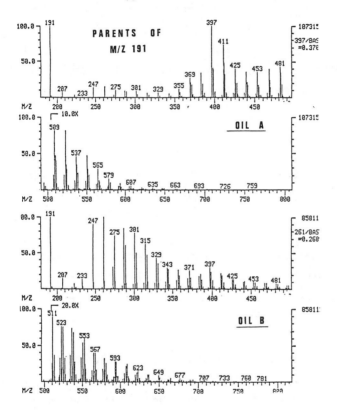

FIGURE 8.   The parent ions of m/z 191 for two oils are shown and it is clear that the distributions are quite different indicating that the oils are not related. The most noticeable difference is the absence of the tricyclic terpanes in oil A.   These analyses were completed under chemical ionization conditions and hence the distribution is of M-1 ions not M.

$C_nH_{(2n-8)}$ are shown. The same technique is applicable to monitoring only the parent ions corresponding to the tri- or tetracyclic compounds.

A major advantage of the direct probe technique is the extension of the carbon number range over which analytes may be characterized. The maximum carbon number normally observed for the hopanes in oils and source rock extracts is in the $C_{35}-C_{40}$ range. Preliminary results from these MS/MS studies suggest that the hopanes extend above C_{40} and possibly as far as C_{70} (Fig. 8a). Continued investigations of geochemical samples by MS/MS will inevitably lead to the discovery of many new high molecular weight biomarkers.

2.1.4 Steranes

Steranes have been widely used to provide source, maturity, migration, and biodegradation information and for oil/oil and oil/source rock correlation[1]. Since collision-activated decomposition of sterane molecular ions yields intense daughters of m/z 217, parent scans may be used to determine sterane distributions. Since the major steranes in most crude oils and source rocks are C_{27}, C_{28}, and C_{29} homologs, the parent scans need only cover a limited range[9,10].

Although direct probe sample introduction does not allow resolution of isomers of a given carbon number, parent scans for the steranes does allow determination of the relative distribution of C_{27} through C_{29} homologs for correlation purposes and to obtain information on the nature of the source material. Marine oils contain a predominance of C_{27} steranes and terrigenous oils are dominated by C_{29} steranes. The parent spectrum of m/z 217

resolution of the diastereomers.

The use of direct probe sample introduction with tandem MS does not permit resolution of diastereomers; however, MS/MS does allow rapid screening of unfractionated samples with separation of tricyclics, tetracyclics, and pentacyclics above C_{45}. Oils and extracts possessing similar terpane distributions would be suspected of originating from the same or similar sources, and confirmation can come from examination of other classes of biomarkers (Figs. 8a and b).

To reduce the complexity of the spectrum and improve signal-to-noise ratio, the technique of selected reaction monitoring (SRM) can be employed. Fig. 9 shows all parents of m/z 191 from an unfractionated crude oil sample, and in the top spectrum only the parent ions corresponding to the pentacyclic hopane molecular ions of molecular formula

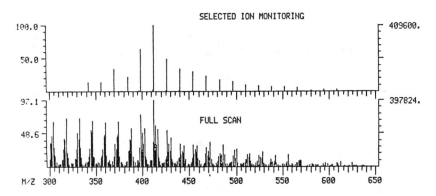

FIGURE 9a and b. The use of selected ion monitoring for the parent ions greatly simplifies the parent ion distributions that are obtained. This figure compares the parent ions of m/z 191 obtained in the full scan mode with those obtained by selected ion monitoring of the parents of the pentacyclic terpanes.

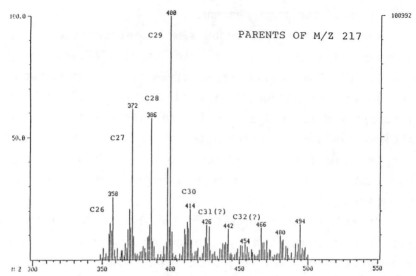

FIGURE 10.   Sterane distributions can be obtained by monitoring the parents of m/z 217. The abundance of the C_{29} steranes and the probable presence of C_{26}, C_{30}, C_{31} and C_{32} steranes in addition to the $C_{27}-C_{29}$ steranes can be seen in this spectrum.

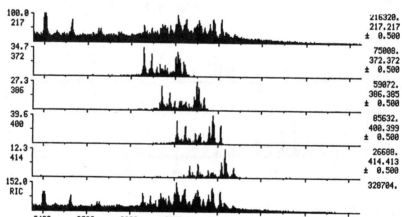

FIGURE 11.   Utilization of GC-MS/MS permits results to be obtained which are similar to those obtained by metastable ion monitoring. In this example GC-MS/MS has been used to analyse for steranes (m/z 217) and subsequently separate out the isomers for the individual members of the homologous series (i.e. C_{27}, C_{28}, C_{29} and C_{30}).

in Fig. 10 clearly indicates the presence of C_{26} and C_{30} steranes and low amounts of the C_{31} and C_{32} homologs. C_{30} steranes have recently been described by Moldowan[11] as being indicative of marine source material, no previous evidence of C_{31} and C_{32} steranes has been reported based on GC/MS data.

2.2 GC/MS/MS of Biomarkers

Highly specific analyses of especially complex samples can be accomplished by using chromatographic preseparation in conjunction with the MS/MS techniques described above. GC/MS/MS analysis allows direct comparison of MS/MS results with GC/MS data, and removes interferences to which GC/MS using single ion monitoring techniques is susceptible.

A paper by Warburton and Zumberge[12] described the use of gas chromatography/metastable ion monitoring using linked scans on a magnetic sector mass spectrometer to determine the distribution of specific biomarkers in complex mixtures. By using collisional activation on the triple-stage quadrupole, a greater degree of control over the decomposition of ions is possible (by varying the collision gas pressure and collision energy), and the spectra are acquired with unit mass resolution. The use of a sophisticated instrument control system allows such experiments to be performed rapidly enough to obtain several different kinds of data from a single injection.

For example, the isomer-specific distribution of steranes illustrated in Fig. 11 was determined by capillary GC/MS/MS. The masses corresponding to the molecular ions that fragment to produce the characteristic sterane daughter ion at m/z 217 are plotted against time. The isomer distributions of C_{27}, C_{28}, C_{29}, and C_{30} steranes (m/z 372, 386, 400, 414, respectively) are well resolved by the chromatographic column. The two stages of mass analysis involved serve to effectively remove interferences from any other compounds present. The end result is a set of well-resolved stereoisomer distributions for each member of the sterane series, which greatly facilitates determination of maturity and other parameters that require isomer-specific data.

3. CONCLUSIONS

Data presented in this paper have demonstrated the use of MS/MS as an alternative method for determining biomarker distributions in fossil fuel samples. The major disadvantage of the technique at this time is its inability to resolve the complex mixtures of stereoisomers present in many geochemical samples. However this disadvantage is outweighed by many advantages of using MS/MS as a screening tool and providing preliminary data on source material and in some cases, maturity data.

The use of the direct insertion probe for analysis of unfractionated oils of extracts provides a rapid method for determining distributions of several classes of biomarkers over an extended carbon number range from a single sample analysis. The use of GC with MS/MS allows a heretofore unavailable degree of specificity. It is clear that MS/MS can play an extremely important role in rapidly screening large numbers of geochemical samples for significant compounds. The information obtained from MS/MS analyses can then be used to indicate those samples that are candidates for further analysis by GC/MS or GC/MS/MS.

REFERENCES

1. R. P. PHILP, Fossil Fuel Biomarkers - Applications and Spectra (Elsevier, Amsterdam, 1985), 294 p.
2. A. S. MACKENZIE, in Advances in Petroleum Geochemistry Vol. 1, edited by J. Books and D. Welte (Academic Press, London, 1984).
3. W. K. SEIFERT and J. M. MOLDOWAN, Geochim. Cosmochim. Acta, 42, 77 (1978)
4. W. K. SEIFERT, J. M. MOLDOWAN, and R. W. JONES, Proc. 10th World Petroleum Congr., SP8, p. 425, 1979. (Heyden, London).
5. J. VOLKMAN, R. ALEXANDER, R. I. KAGI, R. A. NOBLE and G. W. WOODHOUSE, Geochim. Cosmochim. Acta, 47, 2091 (1983)
6. E. J. GALLEGOS, Anal. Chem., 48, 1348 (1976).
7. R. P. PHILP, T. D. GILBERT, and J. FRIEDRICH, Geochim. Cosmochim. Acta, 45, 1173 (1981)
8. J. M. MOLDOWAN, W. K. SEIFERT, and E. J. GALLEGOS, Geochim. Cosmochim. Acta, 47, 1531 (1983)
9. A. S. MACKENZIE, S. C. BRASSELL, G. EGLINTON, and J. R. MAXWELL, Science, 217, 491 (1982)
10. W. K. SEIFERT and J. M. MOLDOWAN, Geochim. Cosmochim. Acta, 45, 783 (1981)
11. J. M. MOLDOWAN, Geochim. Cosmochim. Acta, 48, 2767 (1984).
12. G. A. WARBURTON and J. E. ZUMBERGE, Anal. Chem., 55, 123 (1983)

SOME FEATURES OF PORPHYRINS AND OTHER BIOMARKERS IN
CRUDE OIL AND SOURCE ROCK FROM CONTINENTAL SALT-LAKE
SEDIMENTS IN CHINA

ZHIQIONG YANG
YUYING TONG
Beijing Research Institute of Petroleum
Exploration and Development
P.O. Box 910
Beijing, People's Republic of China

ZHAOAN FAN
Institute of Petroleum
Exploration and Development
Jianghan, Hubei Province,
People's Republic of China

Porphyrins and other biomarkers in crude oils and
source rocks from several basins, especially the
Jianghan Salt-lake Basin in China, are discussed. It
is concluded that the main types of free porphyrin from
continental sediments are similar to those from marine
sediments, except that only nickel porphyrins exist
in the contentinental sediments. A large amount of
gammacerane is a characteristic feature of the
continental salt-lake sediments of Jianghan. In
addition, it is suggested that tricyclic terpane/total
terpane (%), $C_{27} + C_{28}$/gammacerane, gammacerane/17 α(H)
hopane ratios, and changes in nickel porphyrin content
can be considered as indicators of maturity.

1. INTRODUCTION

Since the first study of porphyrin compounds from continental crude oils in the early 1960's, the authors have analyzed
more than 100 samples of source rocks and crude oils from

Jianghan, Biyang, Jizhon and Tsaidam Basins.[1] This paper
deals with the Jianghan Basin which is a typical Ceno-Meso-
zoic continental salt-lake basin. Its Lower Tertiary source
rocks are very unusual due to their high salinity and
strongly reductive depositional environment.[2] Several problems have been investigated, biomarkers such as porphyrins,
perylenes, terpanes and steranes, plus data on nickel and
vanadium elements in the crude oils. These problems include
sedimentary environment and the evolution and correlation
of source rocks and crude oils.

2. EXPERIMENTAL CONDITIONS

2.1. Samples

A total of 39 crude oils and 25 rock samples was obtained
from Tertiary formations, plus one oil sample from Qu-1.
Quantitative and qualitative analyses of the porphyrin compounds were carried out. Demetallated porphyrins in oils
from the Jianghan Basin were identified by mass spectrometry.
Seven crude oils and five source rocks from Qianjiang and
Xingouzui formations in the Jianghan Basin and Qu-1 well
in Tarim Basin were analyzed by GC-MS. The five oils
mentioned above were directly analyzed by infrared spectroscopy without pretreatment and their extracts of saturates,
aromatics, non-hydrocarbons and asphaltenes were also analyzed by infrared spectroscopy. In addition they were ashed
and their trace element contents examined by emission
spectroscopy.

2.2. Sample Processing

After asphaltenes were precipitated from oil samples, aromatic and saturated hydrocarbons were separated by
chromatographic columns. The saturated hydrocarbons were

identified by GC–MS and the aromatic hydrocarbons were
identified by UV and Fluorescence spectroscopy. A portion
of the aromatics containing a high porphyrin concentration
was first treated with methyl sulfonic acid for demetal‍
lation, purified by passage through chromatographic columns,
and then identified by MS. Source rock samples were extracted with ethanlo–benzene solvent (1:9), and the extract
processed in a similar manner.

   Instruments: Specord UV. VIS.

         Jp–3D Fluorescence spectrophotometer

         P–E577 Infrared spectrophotometer

         JGC–2OKP/JMS–D300/JMA–2000

  Experimental conditions for the analysis of terpanes
and steranes by GC–MS are as follows:

  GC: OV–101 capillary chromatography column, with an
internal diameter of 0.5nm and a length of 45m; temperature
program 160–280°C with a rate of 4°C/min; injection temperature 300°C, using He carrier gas at a flow rate of 1 ml/
min.

  MS: electronic bombardment ion source (EI), ionization
voltage, 70 eV, ionization current, 300 μA, resolution, R=
500 interfacial temperature 260°C. Scanning controlled by
computer, scanning speed, 4 seconds, mass range, 56–540,
recording time, 100 min.

3. EXPERIMENTAL RESULTS

3.1. Porphyrin Type and Content

Only nickel porphyrins were found in the source rocks and
crude oils from Jianghan Basin. The wavelengths of the absorption peaks were 514 and 553nm (Fig. 1). Metal porphyrins were demetalled with methylsulfonic acid to form free
base porphyrins. Among the four major absorption peaks, the

fourth one (I_4, 500nm) is the strongest and the third (I_3, 534nm) is the weakest, with the ratio $I_4/I_3 = 3.8$ being characteristic of the porphyrin mixtures. The first peak (617 nm) is weaker than the second one (564nm), and some samples show a weak absorption peak near 600nm, all of which appear to be characteristic of the ETIO type[6].

After purification by silica gel chromatography, MS analysis of the free porphyrins shows that crude oils from the Jianghan Basin are mainly of the DPEP and ETIO series with a small amount of Di-DPEP and RHODO series. Molecular weights of the DPEP and ETIO series are distributed through a wide range (Table 1). MS results coincide with those data obtained from the visible spectrum. Using the height of the absorption peak at 553nm,

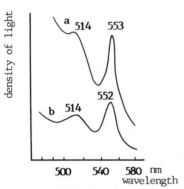

FIGURE 1 UV-vis. absorptive
spectrum of porphyrins
from oils
a: oil from Guang 33 well
b: source rocks from Guang
27 well

the amounts of porphyrins were calculated by Buger-Bland-Beer's law. The average amounts of porphyrins from Qian-jaing formation and lower part of Xing-1 section are listed in Table 2.

In general, porphyrin content decreases with increasing depth of burial and maturity of the crude oil. Table 2 indicates that the porphyrin content is highest in the Qian-1 section. The porphyrin content of the Qian-2 section is somewhat lower than that of the Qian-3 section, possibly because of the higher salt content in this section.

3.2. MS Characteristics of Terpanes and Steranes

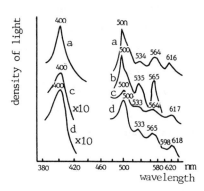

FIGURE 2 UV—vis absorptive spectrum
of demetallated porphy-
rins from crude oils.
(a) standard sample (Prof. YEN)
(b) acetic—KBR from Ying—2 well in
    Shengli oil field, China.
(c) methanesulfonic acid from
    Wang—13 well in Jianghan oil
    field , China.
(d) methanesulfonic acid from
    Wang 1—34—9 well in Jianghan
    oil field, China

TABLE 1   Free porphyrins data for crude oils

Alkyl carbon number	carbon number	molecular weight			
		DPEP (308+14n)	ETIO (310+14n)	DI—DPEP (358+14n)	RHODO (306+14n)
7	27	406		404	
8	28	420	422	418	
9	29	434	436	432	484
10	30	448	450	446	498
11	31	462	464	460	512
12	32	476	478	474	526
13	33	490	492	488	
14	34	504	506	502	
15	35	518	520	516	
16	36	532	534		

TABLE 2   Average porphyrin data from oils and rocks

section	crude					oil				rock samples
	number of samples	s.g.	visc (cp)	saturates (%)	Asphaltenes $\frac{1600cm^{-1}}{1460cm^{-1}}$	Ni—porphyrins (ppm)	V (ppm)	Ni (ppm)	Ni/V	Ni—porphyrins/number of samples (ppb)
Q—1	2	0.9403	195.0	19.64	0.37	133.2	2.38	>15.2		152.2/2
Q—2	3	0.9102	115.3	32.02	0.42	34.2	1.01	12.34	12.22	29.6/3
Q—3	12	0.8846	62.5	37.74	0.54	42.5	1.07	10.38	9.70	74.4/6
Q—4	13	0.8466	11.9	57.54	0.60	11.7	0.48	4.34	9.04	18.7/7
X—1	8			66.43		5.6	0.14	0.94	6.71	no/2

ZHIQIONG YANG, YUYING TONG, ZHAOAN FAN

The GC-MS analyses of saturated hydrocarbons in eleven oils and source rock samples from the Jianghan Basin show that tricyclic diterpanes, pentacyclic triterpanes, and steranes are quite abundant. A sample from Huang well No. 18 is taken as an example (Fig. 3-1 and 3-2).

Figure 3-1 is a m/e 101 mass chromatogram showing an abundance of alkyl tricyclic diterpanes (peaks 3-10, 1 and 2 not shown).

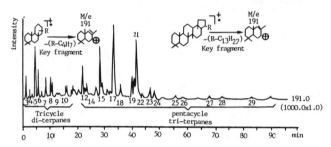

FIGURE 3-1   Distribution of terpanes in Huang-18 crude oil
from Jianghan Basin

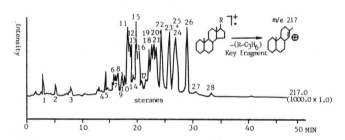

FIGURE 3-2   Distribution of steranes in Huang-18 crude oil
from Jianghan Basin

Their identification is mainly based on mass spectra plus the mass chromatograms of m/e191, 262, 276...388. Identification of pentacyclic triterpanes is based on their mass spectra plus the m/e191, 370, 384...482 mass chromatograms (Table 3). The prominent peak No. 21 in Fig. 3-1 corresponds to gamma cerane ($C_{30}H_{52}$). Its fragmentation

TABLE 3   Terpanes from oils and source rocks in Jianghan

* peak No.	mol. formula	mol. wt.	name of biomarker	basis+
1	$C_{19}H_{34}$	262	tricyclic diterpane	MC
2	$C_{20}H_{36}$	276	tricyclic diterpane	MC
3	$C_{21}H_{38}$	290	extended tricyclic diterpane	MC Bar G.
4	$C_{22}H_{40}$	304	extended tricyclic diterpane	MC
5	$C_{23}H_{42}$	318	extended tricyclic diterpane	MC Bar G.
6	$C_{24}H_{44}$	332	extended tricyclic diterpane	MC
7	$C_{25}H_{46}$	346	extended tricyclic diterpane	MC
8	$C_{26}H_{48}$	360	extended tricyclic diterpane	MC
9	$C_{27}H_{50}$	374	extended tricyclic diterpane	MC
10	$C_{28}H_{52}$	388	extended tricyclic diterpane	MC
11	$C_{27}H_{46}$	370	$18\alpha(H)$–22,29,30–Trisnorhopane	MC
12	$C_{30}H_{52}$	414	Onocerane	MC
13	$C_{27}H_{46}$	370	$17\alpha(H)$–22,29,30–Trisnorhopane	MC
14	$C_{28}H_{48}$	384	Pentacyclic triterpane	MC
15	$C_{29}H_{50}$	398	$17\alpha(H),21\beta(H)$–30–Norhopane	MC
16	$C_{29}H_{50}$	398	Pentacyclic triterpane	MC
17	$C_{30}H_{52}$	412	$17\alpha(H),21\beta(H)$–Hopane	MC Bar G.
18	$C_{30}H_{52}$	412	$17\beta(H),21\alpha(H)$–Moretane	MC
19	$C_{31}H_{54}$	426	$22S$–$17\alpha(H),21\beta(H)$–30–Homohopane	MC
20	$C_{31}H_{54}$	426	$22R$–$17\alpha(H),21\beta(H)$–30–Homohopane	MC
21	$C_{30}H_{52}$	412	Gammacerane	MC Bar G.
22	$C_{31}H_{54}$	426	$17\beta(H),21\alpha(H)$–30–homomoretane	MC
23	$C_{32}H_{56}$	440	$22S$–$17\alpha(H),21\beta(H)$ –30,31–Bishomohopane	MC
24	$C_{32}H_{56}$	440	$22R$–$17\alpha(H),21\beta(H)$–30,31 –Bishomohopane	MC
25	$C_{33}H_{58}$	454	$22S$–$17\alpha(H),21\beta(H)$–30,31,32 –Trishomohopane	MC
26	$C_{33}H_{58}$	454	$22R$–$17\alpha(H),21\beta(H)$–30,31,32 –Trishomohopane	MC
27	$C_{34}H_{60}$	468	$22S$–$17\alpha(H),21\beta(H)$–30,31,32,33 –Tetrakishomohopane	MC
28	$C_{34}H_{60}$	468	$22R$–$17\alpha(H),21\beta(H)$–30,31,32,33 –Tetrakishomohopane	MC
29	$C_{35}H_{62}$	482	$22S$–$17\alpha(H),21\beta(H)$–30,31,32,33,34 –Pentakishomohopane	MC
30	$C_{35}H_{62}$	482	$22R$–$17\alpha(H),21\beta(H)$–30,31,32,33,34 –Pentakishomohopane	MC

* The peak No. with reference to Figure 3–1

+ MC–mass chromatogram. Bar G —Bar graphic.

Zhiqiong Yang, Yuying Tong, Zhaoan Fan

TABLE 4 Steranes from oils and source rocks in Jianghan Basin

[*]peak No.	mol. formular	mol. wt.	name of biomarker		basis[+]
1	$C_{21}H_{36}$	28.	Pregnane		MC Bar G.
2	$C_{22}H_{38}$	302	Sterane		MC
3	$C_{23}H_{40}$	316	Sterane		MC
4	$C_{27}H_{48}$	372	$13\beta,17\alpha$-Diacholestane	20S	MC Bar G.
5	$C_{27}H_{48}$	372	$13\beta,17\alpha$-Diacholestane	20R	MC
6	$C_{27}H_{48}$	372	$13\alpha,17\beta$-Diacholestane	20S	MC
7	$C_{28}H_{50}$	386	Diacholestane		MC
8	$C_{27}H_{48}$	372	$13\alpha,17\beta$-Diacholestane	20R	MC
9	$C_{28}H_{50}$	386	24-Methy1-$13\beta,17\alpha$ -diacholestane	20S	MC
10	$C_{28}H_{50}$	386	24-Methy1-$13\beta,17\alpha$ -diacholestane	20R	MC
11	$C_{27}H_{48}$	372	$5\alpha,14\alpha,17\alpha$-Cholestane	20S	MC Bar G.
12	$C_{27}H_{48}$	372	$5\beta,14\alpha,17\alpha$-Coprostane		MC Bar G.
13	$C_{27}H_{48}$	372	$5\alpha,14\beta,17\beta$-isocholestane	20S	MC
14	$C_{28}H_{50}$	386	24-Methy1-$13\alpha,17\beta$-Diachostane		MC Bar G.
15	$C_{27}H_{48}$	372	$5\alpha,14\alpha,17\alpha$-Cholestane	20R	MC
16	$C_{29}H_{52}$	400	24-Ethy1-$13\alpha,17\beta$ -diacholestane		MC
17	$C_{29}H_{52}$	400	24-Ethy1-$13\alpha,17\beta$ -diacholestane		MC
18	$C_{28}H_{50}$	386	24-Methy1-$5\alpha,14\alpha,17\alpha$ -cholestane	20S	MC
19	$C_{28}H_{50}$	386	24-Methy1-$5\beta,14\alpha,17\alpha$ coprostane	20R	MC
20	$C_{28}H_{50}$	386	24-Methy1-$5\alpha,14\beta,17\beta$ -isocholestane	20S	MC
21	$C_{28}H_{50}$	386	24-Methy1-$5\alpha,14\beta,17\beta$ isocholestane	20R	MC
22	$C_{28}H_{52}$	386	24-Methy1-$5\alpha,14\alpha,17\alpha$ -cholestane	20R	MC
23	$C_{29}H_{50}$	400	24-Ethy1-$5\alpha,14\alpha,17\alpha$ -cholestane	20S	MC
24	$C_{29}H_{52}$	400	24-Ethy1-$5\alpha,14\beta,17\beta$ -isocholestane		MC
25	$C_{29}H_{52}$	400	24-Ethy1-$5\beta,14\alpha,17\alpha$ -coprostane		MC
26	$C_{29}H_{52}$	400	24-Ethy1-$5\alpha,14\alpha,17\alpha$ -cholestane	20R	MC
27	$C_{30}H_{54}$	414	4-Methy1-24-ethy1-cholestane		MC
28	$C_{30}H_{54}$	414	4-Methy1-24-ethy-cholestane		MC

[*] The peak No. with reference to Figure 3-2
[+] MC – mass chromatogram   Bar G.–Bar grapnic(mass spectrun)

pattern (Fig. 4) and retention time are slightly different
from those of 17 α (H) hopane.

Figure 3-2 is a m/e 217 mass chromatogram in which
numbered peaks correspond to compounds listed in Table 4.
Rearranged sterane peaks 4-10, 14, 16, 17 are identified
mainly by the diagnostic fragment m/e 259 (Fig. 5). Peaks
27 and 28 are 4-methyl-24-ethyl cholestane, being distinguished by the diagnostic fragment m/e 231. Pregnane (peak
No. 1) commonly appears in the samples from the Jianghan
Basin, and it is easily identified by the m/e 288 mass
chromatogram and by its mass spectrum shown in Fig. 6.

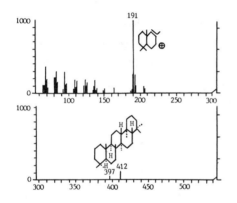

FIGURE 4

Mass spectrum of
Gammacerane

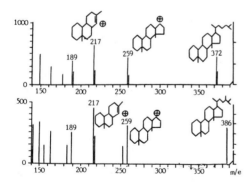

FIGURE 5

Mass spectra of
diacholestane and
diaergostane
(rearranged steranes)

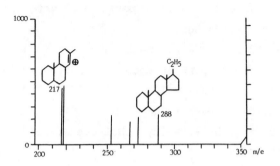

FIGURE   6   Mass spectrum of pregnane

3.3   Perylenes

Perylenes were found in the aromatic hydrocarbons of three
rock samples from Wang well 1-34-9 (1314m depth), from
Zhong well 33 (1978m depth), and from Zhong 29 (1489.5m depth)
Fig. 7 and 8.   Perylene is a stable pentacyclic aromatic compound originating from terrigenous organisms, and its presence proves that continental deposits in the Jianghan
Basin contain some terrigenous materials.

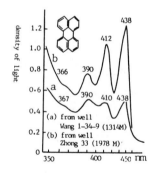

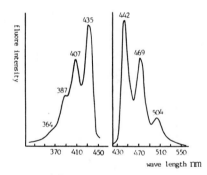

FIGURE 7 UV-vis spectrum for
perylenes from rock
extracts

FIGURE 8 (1) Excitation spectrum
for perylene from rock
Wang well 1-34-9 (1214m)

(2) Fluorescence curve for
perylene from rock, Wang
well 1-34-9 (1314 m)

4. DISCUSSION

4.1. Dipositional Environment and Types of Organic Matter

Porphyrins from marine oils are usually predominantly
vanadium, but in Chinese continental oils, nickel porphyrins
predominate. Samples from Jianghan Basin show only
nickel porphyrins. Furthermore, in the combustion analysis
of the Jianghan Basin oils, nickel content strongly pre-
dominates over vanadium, indicating that crude oils from
that basin are of a typical continental origin (Table 2).

Pregnane ($C_{21}H_{36}$) is found in most of the samples and
is relatively abundant in both Qian-4 and Xing-1 sections,
unlike the Carboniferous strata in Tarim Basin Xingjiang.
There has been little published data on pregnane and its
origin should be investigated in future work.

Abundant gammacerane, which possibly originates from
protozoans,and perylenes, which are thought to originate
from terrigenous plants, were found in the organic matter
from the Qianjiang formation. This coincides with the
mixed type kerogens found therein.

Although gammacerane occurs in other districts,[3] its
abundance is not as high as that in Qianjiang formation of
the Jianghan Basin. Rock extracts from Wang well 4-11-3
have a gammacerane content which is even higher than $17\alpha(H)$
hopane, distinguishing them from marine oils. It could be
suggested that high gammacerane content is a characteristic
of some continental oils, but its relation with the salt-
lake environment needs further study.

4.2. Maturity of Crude Oil and Source Rocks

Porphyrin content determined for oils from the Jianghan
Basin follows a trend as seen in Table 2. It is highest

in oils of the Qian-1 section showing that they are very immature. Porphyrin content of oils in Qian-4 section and the
lower part of Xing-1 section is low and even absent in some
cases, indicating a higher maturity. These results correspond to other data in Tables 2 and 5. The depth of burial
of the Qianjiang formation is 2800 to 3000 meters and its
temperature is between $110-120^{\circ}C$. Porphyrins may be destroyed under these conditions having no porphyrins remaining. Therefore, porphyrin concentrations can be used as
indicators of oil maturity. Table 5 lists biomarker parameters from GC-MS analysis plus several conventional characteristics of oils from different depth intervals. These
show that the maturity of the oils increases gradually from
Qian-1 to Qian-4 section. Although the depth of burial of
oils from the lower part of Xing-1 section is shallower,
their maturity is high. Comparison of the crude oil from
Guang well 4-2 show the former to be more mature than the
latter. This is related with depth of burial.

Pentacyclic triterpanes (C_{29}, C_{28}) are labelled as secondary[4], thus, the ratio of $C_{29} + C_{28}/C_{29} + C_{30}$ will be
greater with increasing maturity.. The Σtricyclic terpane/
total terpane (%) ratio has also been taken as an indicator
of oil maturity[4]. Ratios of gammacerane/17α(H) hopane and
pentacyclic triterpanes ($C_{27} + C_{28}$)/gammacerane also very directly with maturity as shown by the trends in Table 5. In
considering the homohopanes (C_{31}), the 22R stereoisomer will
transform into 22S with increasing maturity[5]. The data in
Table 5 shows that rearranged steranes total steranes (%)
also increases with maturity. In recent sediments, steranes
exist as 20R changing into 20S as depth and maturity increase. This is corroborated by the data in Table 5.

The maturity of organic matter in rock samples is main-

TABLE 5 Maturity correlation of oils and source rocks

	well No.	depth(m)	strato	saturates (%)	main C	OEP	Pr/Ph	porphyrin (ppm)* (ppb)	$\frac{C_{27}+C_{28}}{C_{29}+C_{30}}$	tricyclic % total terpane	$\frac{C_{31}-22S}{C_{31}-22R}$	$\frac{C_{27}+C_{28}}{\text{gammacerane}}$	$\frac{\text{gammacerane}}{17(H)\text{hopane}}$	$\frac{C_{29}\ 20S}{C_{29}\ 20R}$	rearranged sterane/total sterane %
oil	W 1-34-9	1254.6-1293.6	Eq1	9.50	C_{28}	0.32	0.0638	101.6	0.06	0	1.05	0.09	2.33	0.54	12.0
	G 33	1829.3-1829.6	Eq1	29.77	C_{17}	0.78	0.2478	164.8	0.10	18.4	0.78	0.29	0.83	0.41	5.7
	W 4-2	1307.58-1309.46	Eq2	27.11	C_{18}	0.70	0.1913	22.7	0.13	14.4	1.50	0.38	0.88	0.91	13.3
crude	G 3-11	2827.2-2857.0	Eq3	48.62	C_{22}	0.82	0.3316	12.2	0.17	22.1	1.14	0.76	0.50	0.76	15.1
	H 18	2616.4-2622.8	Eq4	55.59	C_{17}	0.83	0.4470	trace	0.22	34.1	1.26	0.82	0.78	0.95	24.2
	L 43	1114.2-1140.7	Ex1	69.15	C_{19}	0.91	0.7130	no	0.30	58.2	1.15	1.50	0.48	1.18	31.2
source rocks	W 4-11-3	1214.34-1234.9	Eq2	9.8	C_{28}	0.44	0.1098	63.6*	0.17	9.4	0.59	0.37	1.79	0.51	11.1
	G 27	1917	Eq1	15.75	C_{20}	0.92	0.1574	287.2*	0.17	17.8	0.79	0.96	0.56	0.35	9.9
	H 8	2383.5	Eq4	28.69	C_{24}	0.98	0.3920	no*	0.18	8.0	1.25	1.19	0.34	0.68	23.8
	G4-13	2853.5	Eq3	33.63	C_{19}	1.11	0.6693	157.0	0.19		1.52	4.89	0.08	0.93	

ly controlled by temperature which generally increases with depth of burial. Figure 9 shows that in wells Wang 4-11-3 and Guang 27 homohopane stereoisomer 22S is weaker than 22R and 17β (H) moretane and gammacerane are quite rich. Therefore the two samples are of low maturity. For wells Huang 8 and Guang 4-13, the ratio of C_{31} $22S/C_{31}22R$ ranges from 1 to 1.5 and 17β (H) moretane and gammacerane are weaker, indicating the greater maturity of these samples. This is generally supported by other parameters in Table 5.

4.3. Correlation of Crude Oils with Sources

By comparing the porphyrin contents in terpanes and steranes from 5 crude oils with their corresponding source rocks in the Qianjiang formation of the Jianghan Basin, some conclusions can be made:

1. Based on the relative intensities of the peaks of pentacyclic triterpanes (Fig. 10) the crude oils from Qianjiang and Xingouzui formations differ slightly because they come from different source rocks. The five samples from Qianjiang formation are similar and therefore they may be derived from the same oil source. Since the fingerprints of Wang well 4-2 (Qian-2 section) and that of Guang well 3-11 (Qian-3 section) are similar, oil from Wang 4-2 may come from Qian-3 section. The oil from Huang well 18) (Qian-4 section) shows a slight difference, indicating that it may have been generated and remained in place.

2. Fig. 11 indicates a good relationship between porphyrin content of crude oils and source rocks. For Qian-3 Qian-4 and Xin-gouzui formation, source rocks are mature and are capable of supplying oils for their corresponding reservoirs.

3. Oil from wells Wang 1-34-9 and Wang 4-2 come from

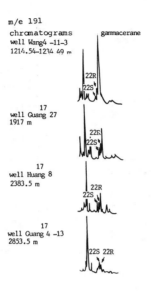

FIGURE 9 Thermal evolution of pentacyclic triterpane stereoisomers
from source rocks of
Jianghan Basin

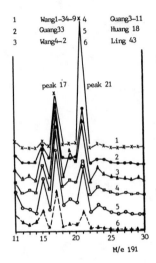

1	Wang1–34–9	× 4	Guang3–11
2	Guang33	5	Huang 18
3	Wang4–2	6	Ling 43

peak 17   peak 21

M/e 191

FIGURE 10 Fingerprints of
of crude oils. peak 17–17α(H),
21α(H)-hopane
peak 21 – gammacerane
1, 2, 3, 4, 5-Qianjiang
6-Xingouzui

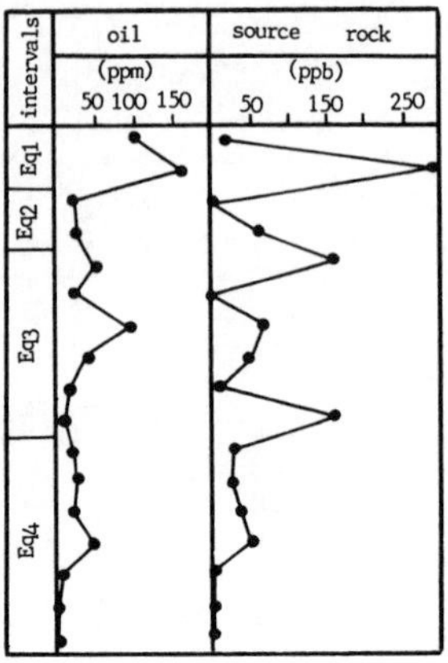

FIGURE 11 Correlation of
porphyrin content for oils
and source rocks in
Qianjiang formation

shallower depths and the source rocks are not fully mature. Those crude oils show characteristics of low maturity and were considered to have been formed in place. As the source rocks in Qian-3 and Qian-4 sections are very mature, the crude oils (especially Qian-3 section) may be assumed to have migrated into Qian-1 and Qian-2 reservoirs through faults and porous sandstone. Therefore, the crude oils of Qian-1 and Qian-2 sections may come from a combined source i.e. from the source rocks in the vicinity and from Qian-3 and Qian-4 sections.

CONCLUSIONS

1. Nickel porphyrin content of oils decreases with maturity, and its content in the rock samples change with depth.

2. In the Jianghan Basin free porphyrins are predominantly DPEP and ETIO series with a minor amount of Di-DPEP and RHODO series.

3. Oils from the Jianghan Basin show a continental origin, and kerogen in the Qianjiang formation is of a mixed type. The presence of a large amount of gammacerane appears characteristic of continental deposition.

4. Maturity in Qian-1 and Qian-2 intervals is low, while that of Qian-3 and Qian-4 intervals is high. Although the depth of burial of the lower part of the Xin-1 interval is shallower, its degree of maturity is the highest. Ratios of Σ tricyclic terpane/total terpane (%), $C_{27}+C_{28}$/gammacerane, and gammacerane/17α(H) hopane can be used as indicators of maturity, showing a good correlation with other indicators.

5. It is likely that oil originating from the Qianjiang formation is different from the Xingouzui formation. Each interval of Qianjiang formation seems to have the character-

istics of acting as an oil source and reservoir. Oil from
the Qian-2 interval of Wang well 4-2 probably comes from
the Qian-3 interval.

REFERENCES

1. Yang Z., Sun, C.Z., Report on petroleum Explora-
    tion Research Vol. 2 (China Industry Pub., Beijing
    1963)
2. Jiang, J., Zhang, Q., Oil & Gas Geology (China), 3,
    1-15 (1982).
3. Shi, J., Geochimica (China) 1, 5-19 (1982).
4. Seifert, W.K., Moldwan, J.M., Geochim. Cosmochim.
    Acta., 42, 77-95 (1978).
5. Seifert, W.K., Moldowan, J.M., In Advances in
    Organic Organic Geochemistry, 1979 (A.G. Douglas
    and J.R. Maxwell, eds.) pp 229-237, Pergamon Press
    (1980).
6. Baker, E.W., Yen, T.F., Amer. Chem. Soc., 89, p.
    3631 (1967).

ANALYSIS OF CHLOROPHYLL DIAGENESIS III. THE EFFECT OF METALS ON THE PHORBIDE TO PORPHYRIN TRANSITION

Pamela P. Zelmer, Jumat Bin Salimon, and
Eugene H. Man

Department of Chemistry
University of Miami
Coral Gables, FL   33124

Compounds related in structure to chlorophyll, especially deoxophylloerythro-etioporphyrin (DPEP) and etioporphyrin, have long been used by the petroleum industry as indicators of thermal maturity. Earlier chlorophyll diagenetic products, especially 7-ethyl-7-despropio deoxomesopyropheophor-bide, have been shown to form DPEP by an aromatization reaction that is time and temperature dependent (4). Sedimentary analyses suggest that this reaction takes place between the free base forms of the chlorophyll derivatives, followed by a metalation reaction with nickel to form nickel(II) DPEP.

We are using laboratory simulation studies to quantify these reactions in order to determine the parameters needed for developing a thermal modeling equation. Our studies have led us to conclude that in spite of what appears to be a free base reaction in sediments, that the reaction takes place by initial formation of the metallo-derivative of the phorbide, 7-ethyl-7-despropio deoxo-mesopyropheophorbide, which rapidly aromatizes to form metallo-DPEP. The evidence for proposing this pathway for the formation of Ni(II)DPEP is presented in this report.

INTRODUCTION

Chlorophyll is the most abundant organic compound on earth (1) being found in all terrestrial green plants, and in the blue green algae of the oceans which form the basis of the oceanic food chains. Because it is the earth's primary agent of photosynthesis, all life depends on it.

The basis of the chlorophyll molecule is a highly aromatic tetrapyrrole (Fig. 1) known as a porphyrin. Hemoglobin, the agent of respiration in most animals, is a fully aromatized porphyrin. Chlorophyll, having one less double bond than hemoglobin, belongs to the subclass phorbide. Chlorophyll also contains an isocyclic ring attached to the C-ring of the tetrapyrrole unit which makes it distinct. No other porphyrin compound in nature contains this isocyclic ring.

Porphyrins chelate with over two-thirds of the metals of the periodic chart. Chlorophyll contains a magnesium ion in its

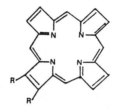

(a) Chlorophyll a

(b) Chlorin nucleus

(c) Porphyrin nucleus

(d) DPEP series

(e) Etio series

Fig. 1. Some important structures in chlorophyll diagenesis

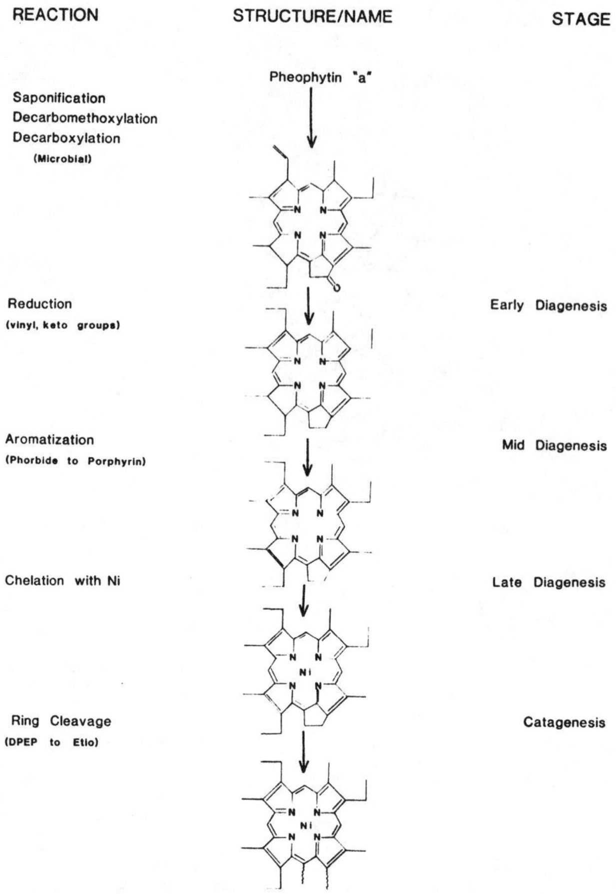

Pamela P. Zelmer

REACTION   STRUCTURE/NAME   STAGE

Pheophytin "a"

Saponification
Decarbomethoxylation
Decarboxylation
(Microbial)

Reduction   Early Diagenesis
(vinyl, keto groups)

Aromatization   Mid Diagenesis
(Phorbide to Porphyrin)

Chelation with Ni   Late Diagenesis

Ring Cleavage   Catagenesis
(DPEP to Etio)

Figure 2   Diagenesis of chlorophyll to nickel
            petroporphyrins based on the work
            of Treibs, Baker and Louda.

center while hemoglobin utilizes iron.   When
not chelated with a metal ion, the porphyrin
nucleus contains two acidic hydrogens and is
known as the "free base" form.

Metalloporphyrins are relatively stable
in geological environments.   Rocks of up to
60 million years of age, as well as petroleum
have been found to contain metalloporphyrins.
The vast majority of these porphyrins are
related to two compounds, deoxophylloerythroetioporphyrin (DPEP) and etioporphyrin
(etio), shown in Figure 1.   Alfred Treibs (2)
was the first to elucidate the structure of
DPEP in petroleum.   Partially because DPEP
contains the same isocyclic ring as found in
chlorophyll, Treibs deduced that DPEP was the
product of chlorophyll diagenesis.   The
diagenetic scheme (Figure 2) which he
postulated to account for the structural
differences between DPEP and chlorophyll
became the basis for the science of
biomarkers, organic compounds found in the
sediment which geochemists use to determine
the history of the sediment.   Biomarkers are
organic compounds buried in the sediment
which undergo a series of progressive
chemical changes; those which involve the
reactions dependent upon both time and
temperature provide the most geochemical
information.

Baker and coworkers (3, 4, 5, 6, 7) have
been responsible for many of the analyses of
chlorophyll degradation products in cores
obtained from the Deep Sea Drilling Project
(DSDP). Their comprehensive analyses have
led to the following conclusions;

1.   The early stages of diagenesis
       occur in shallow sediments where
       conditions are unstable due to high
       microbial activity. Most of the
       chlorophyll is degraded to bile
       pigments by attack on the
       tetrapyrrole unit.

2.   Once the sediment becomes stable
       and slightly anoxic, further
       diagenetic reactions are time and
       temperature dependent.

Sedimentary analyses led Baker et al,
(4, 5) to postulate that the reactions
involved in the second stage of chlorophyll
diagenesis follow the order: (1) aromatization (the phorbide to porphyrin transition);
(2) metalation with nickel; and (3) thermal
degradation (Figure 2). The compounds
involved in these reactions are 7-ethyl-7
despropio deoxomesopyropheophorbide a and
DPEP, both initially in the free base form.
Baker's thermal studies have shown that the
aromatization of a phorbide to the porphyrin

is both time and temperature dependent and
occurs at about 40°C in the sediment.   This
is followed by metalation of the free base
DPEP with nickel at approximately 50°C.

We began to study chlorophyll diagenesis
by measuring the rates of the aromatization
and metalation reactions under laboratory
conditions which simulated those found in
ocean sediments, using deoxomesopyropheophor-
bide a methyl ester (DOMPP) and
desoxophylloerythrin methyl ester (DPE)
(Figure 3) as model compounds.   The carbonyl
groups present on the model compounds, but
absent on the sedimentary compounds, were
masked by forming the methyl ester.

Our studies led us to propose that the
phorbide to porphyrin transition does not
occur via the apparent route deduced from
sedimentary analysis, but that phorbide
undergoes metalation prior to aromatization.
(Figure 4.)   Thus, sedimentary analyses
report the exclusive presence of free base
phorbide because the metalophorbide is a
geologically short-lived species which
aromatizes as quickly as it forms.   Previously published work which led us to propose the
existence of the metalophorbide and the
sedimentary reaction following its formation,
as well as new work supporting this
hypothesis, will be presented in this report.

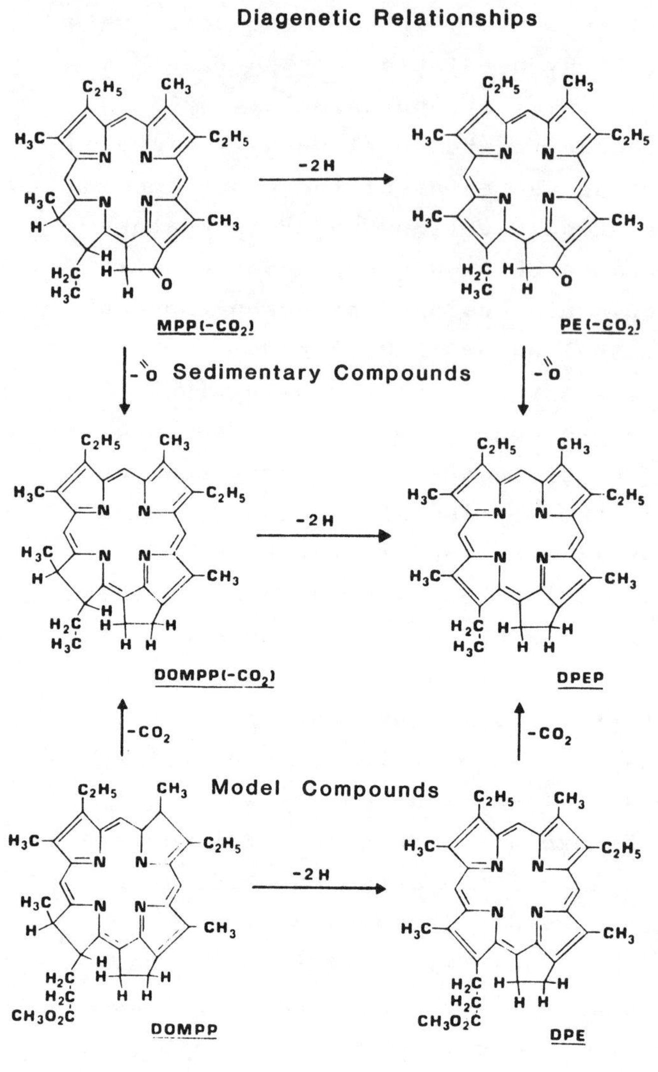

Pamela P. Zelmer

Diagenetic Relationships

Figure 3 Relationships between sedimentary compounds involved in the phorbide to porphyrin transition and those used in the laboratory simulation (see Appendix A for key to abbreviations).

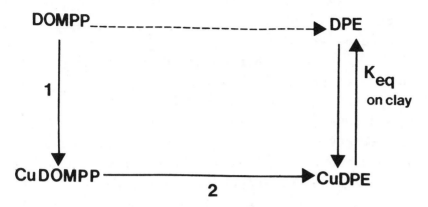

Figure 4   Proposed pathway for conversion of
DOMPP to DPE via metallo-phorbide.

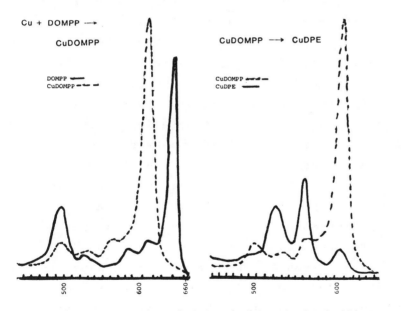

Figure 5   Spectral changes in the phorbide to
porphyrin transition
(Copper(II)complexes).

EXPERIMENTAL

Cu Studies

A portion of this paper includes a summary of previously published work, describing the reaction of Cu with DOMPP to form CuDOMPP, the subsequent aromatization of the CuDOMPP, and the reaction of Cu with DPE to form CuDPE. A detailed explanation of the procedures and syntheses involved can be found in earlier papers (8, 9). The reactions were run under pseudo-first order conditions with the metal in at least fiftyfold excess over the porphyrin. Kinetic analyses of changes in the visible absorption spectra at the appropriate wavelengths (639 nm for DOMPP, 608 nm for CuDOMPP, 620 nm for DPE, and 562 and 525 nm for CuDPE) demonstrated that the reactions were first order in the porphyrin or metalloporphyrin under investigation. The observed rate constants were used to calculate the specific rate constants; all reactions were run at two or more temperatures. Activation parameters were calculated using standard equations.

The reactions were carried out in a series of organic solvents chosen to simulate a variety of sedimentary environments. The solvents used in one or both series discussed below are listed in Table I.

TABLE I

Solvent	Properties	Type Environment
a) THF	nonpolar, aprotic	lipid petroleum fraction
b) DME	nonpolar, aprotic	lipid petroleum fraction
c) diglyme	nonpolar, aprotic	lipid petroleum fraction
d) DMA	polar, aprotic	carbonaceous, protein
e) EtOH	polar, protic	sediment/water interface
f) Mem	polar, protic	sediment/water interface

a)   tetrahydrofuran
b)   dimethoxyethane
c)   bis(2-methoxy)ethane
d)   N,N-dimethylacetamide
e)   ethanol
f)   microemulsion, a toluene, water, isopropanol mixture (Lavalle, 1979)

Not all the solvents were used in both studies discussed.

Metal Series

The insertion of metal ions into DOMPP using THF as solvent was studied to determine the relative ease of insertion of a series of metals. For purposes of comparison, the "reaction times to completion" were determined at a fixed temperature, usually $100\,^{\circ}C$. The appropriate amounts of solid

metal sulfide or aqueous metal ion were added
to DOMPP dissolved in THF. The tubes were
evacuated, sealed and placed in a heating
block. When a color change was observed
(green to blue for metal insertion, blue to
pink for aromatization) the tube was broken
and the visible spectrum recorded. The
spectral changes involved are illustrated in
Figure 5. When the peak under observation
had disappeared, the reaction was considered
to be complete. If the spectrum indicated
that the compound was still present, the
sample was placed in a new tube, resealed and
returned to the heating block. The wavelengths of interest are listed in Table II.

TABLE II

λ Max for Metalation in THF.

	Cu	Fe	Co	Ni
MDOMPP (band 1)	607	608	604	601
(α)	567	569	562	551
MDPE (β)	523	526	520	510

Copper(II) sulfide and nickel(II)
sulfide were purchased from Alpha Products,
Inc. Iron(II) sulfide was obtained from
Fisher Scientific Company. Cobalt(II)
sulfide was synthesized by an adaptation of
the method of Stern (16). All metal sulfides
were purified by Soxhlet extraction in
ethanol. Certified Reagent Grade

(nonsulfide) metal salts: copper(II) nitrate
and nickel(II) nitrate (Fisher Scientific
Company); iron(II) acetate (Alpha Products,
Inc.); and cobalt(II) acetate (Merck and Co.,
Inc.) were purified by recrystallization in
methanol. Tetrahydrofuran was purified by
distillation over sodium and benzoquinone and
used within 4 hours or stored under N2 not
more than 24 hours.

RESULTS AND DISCUSSION
     Initial experiments with the DOMPP to
DPE conversion demonstrated the stability
difference between the compounds involved in
these reactions. When dissolved in THF,
sealed under vacuum, and placed in a heating
block, the resistance to decomposition was in
the order DPE > DOMPP > MDOMPP. Under the
conditions investigated, DOMPP decomposed
rather than aromatized, except in the
presence of a catalyst such as a metal ion.
The metallophorbide formed readily with
various metals under a variety of conditions.
The metallophorbide underwent either
aromatization or decomposition, the relative
rates of which depended upon the environment
and the metal ion.
     The ease with which Cu(II) formed the
metallophorbide, and the similar ease with
which CuDOMPP aromatized to CuDPE led us to
use this metal ion for our initial studies.

Copper phorbides are occasionally found in terrestrial sediments and are used to indicate terrestrial intrusions into deep sea sediments (6). We determined the activation energies for the metalation and aromatization reactions in four organic solvents. The results are listed in Table III.

TABLE III

$$Cu(II) + DOMPP \xrightarrow{-2H^+} CuDOMPP \xrightarrow{-2H\bullet} CuDPE$$
$$(1) \qquad\qquad (2)$$

Solvent	E_{act} (1) Kcal/mole	E_{act} (2) Kcal/mole
DMA	6.00	12.08
EtOH	15.54	12.60
DME	21.30	decomp
THF	19.30	

Cu(II) reacted easily with DOMPP at or near room temperature under all conditions investigated. However, the activation energies varied greatly in the solvents studied. The aromatization reaction was affected by the type of solvent. In the nonpolar DME and THF, decomposition often occurred rather than aromatization. In polar solvents, aromatization proceeded smoothly, with the activation energy apparently unaffected by the differences between solvents.

The relative energies of metalation and aromatization in polar protic ethanol are of

specific interest. In this case, the
activation energy for the insertion of Cu(II)
into DOMPP is slightly higher than the
aromatization of CuCOMPP. Under these
conditions the metalation reaction becomes
the rate-determining step for the overall
conversion of DOMPP to DEP. Thus, sedimentary analyses would not detect the presence
of the metallophorbide since it aromatizes to
form CuDPE as quickly as it is formed.

These reactions were also studied in DMA
where the E_{act} of aromatization is much
higher than the E_{act} of metalation. DMA has
an amide bond which may produce a different
mechanism of insertion from the other
solvents shown in Table III.

The reaction of Cu(II) with DPE was
studied as a function of solvent, and the
results compared to the data obtained from
the reaction of Cu(II) with DOMPP. These two
compounds differ structurally by only one
carbon double bond, but display significantly
different reactivities toward Cu(II). A
comparison of the observed rate constants
extrapolated over the range of 10° to 80°C
showed that under all conditions studied,
DOMPP always reacted faster with CU(II) than
did DPE. The observed rate constants at 25°
and at 40°C as a function of solvent are
listed in Table IV.

TABLE IV

k$_{obs}$ for reaction of Cu(II) with DOMPP and DPE.

Solvent	25° DOMPP/DPE	40° DOMPP/DPE
DOMPP/THF	.081/.028	.389/.073
diglyme	.080/.033	.284/.975
Mem	.083/.051	.280/.148
DMA	.120/.058	.195/.119

The fact that CuDOMPP forms so readily suggests that where DOMPP and CuDPE exist, CuDOMPP will also be found (see Figure 4).

Although Cu(II) is an important metal in terrestrial porphyrins, the metal of interest in deep sea sediments and petroleum is Ni(II). We have expanded our initial model studies of the phorbide to porphyrin transition to include the series of transition metal ions listed in Table V.

TABLE V

$$M + DOMPP \rightarrow MDOMPP \rightarrow MDPE$$
$$M(II) = Co, Ni, Fe, Cu, Mn$$
$$M(III) = Cr, V$$

These metals fall into three distinct groups based on their reactivities. The Group III metals, Cr and V, led to simple decomposition of the phorbide bile pigments. Mn underwent a rapid redox reaction, resulting in total oxidation of the phorbide and reduction of the metal ion to an elemental mirror. The remaining metal ions: Co(II), Ni(II), Fe(II) and Co(II)   formed their respective metallophorbides, which underwent subsequent aromatization to the metalloporphyrins.

We also investigated the effect of the form of the four metals, Co, Ni, Fe, and Cu, in two series of reactions.   In one series, soluble metal ions were dissolved in water before addition to the phorbide solution (free metal ions); in the other, metal sulfides were added (sulfides).   Metals are known to exist in both forms in ocean sediment.   Reaction times to completion in THF at 100° are listed in Table VI for both systems as a function of metal.

TABLE VI

(Hours at 100°C in THF)

	Cu	Fe	Co	Ni
Metalation Reaction				
Sulfide	1/2	13	30	40
Free metal ion	1/6	10	11	18
Aromatization Reaction				
Sulfide	5/2	5	18	10
Free metal ion	2	5	9	6

Three significant results can be seen from these data.   First, the order of

reactivity in the metalation reaction is Cu >
Fe > Co > Ni, regardless of the form of the
metal (sulfide or free metal ion). This is
the same order of reactivity found by
Hambright (11) and Longo (12) in reactions
with tetraphenyl- and etio-type porphyrins,
and would be predicted from the solvent
exchange rates of the metal ions (13). In
the aromatization reaction, the order of
catalytic ability of the metal ions is not
quite the same as the order of reactivities
in the metalation reaction. Here the
stability of the Ni porphyrin may help
increase its reactivity relative to the other
metals.

Second, the effect of the form of the
metal ion (free ion or sulfide) on the
reaction rate and on the relative order of
reactivity in the metalation reaction is much
less than would be predicted, considering the
small effective concentration of the metal
sulfides. The free metal ions, are dissolved
in water before addition to the organic
solvent, are fully solvated in the phorbide
reaction. Thus the concentration of the
metal ion in metal sulfides is up to 10^{-47}
lower than in the free metal ion solutions;
the solubility product of CuS is 10^{-47} (14).
It is apparent from the data that, although
the metal sulfides react more slowly than the
free metal ions, the difference is less than

one power of 10. A possible explanation for
this is that the free base phorbide may
adsorb onto the surface of the sulfide
thereby increasing the effective concentration of the metal ion. Porphyrins have been
shown to adsorb onto a variety of surfaces
and desorb once metalloporphyrin formation
has taken place (15, 8).

Third, the rates of metalation are,
except for Cu, greater than the rates of
aromatization. This result is similar to
that found with Cu(II) and ethanol (8). In
these cases the metallophorbide would be a
short-lived species converting to the
metalloporphyrin in a reaction faster than
the one by which is was formed.

CONCLUSION

The intent of this research was to
investigate the phorbide to porphyrin
transition in a model which simulated
sedimentary environments. Our studies to
date have shown the following to occur in
solution. The kinetics may vary in the
presence of clays, etc.

1.   DOMPP decomposes more readily than
     it aromatizes, except in the
     presence of a catalyst, (Cu(II)).

2.   In polar, protic environments,
     CuDOMPP may aromatize as quickly as
     it forms.

3.   Where CuDPE and DOMPP are present, CuDOMPP must also be present.

4.   Aromatization occurs via the metallophorbide for all metals studied.

5.   For Co, Ni, and Fe in THF, the metallophorbide aromatizes as quickly as it forms.

These data indicate that the sedimentary phorbide to porphyrin transition is catalyzed by metal ions. Although the metallophorbide is, in fact, an intermediate in this transition, it is quickly converted to the metalloporphyrin, and is therefore not found by sediment analyses.

In order for a reaction sequence to be useful for thermal modeling and maturation studies one must know the order of the reactions in the sequence, the rates of the reactions and the effects of the environment on both the order and on the rates. We have shown that the most probable order for the phorbide to porphyrin transition is metalation followed by aromatization, and we have reconciled this with sedimentary data, proposing the pathway shown in Figure 4. We have also shown that the effect of the environment upon this sequence of reactions is great, differs for each reaction, and must be carefully considered when drawing conclusions about biomarkers.

ACKNOWLEDGEMENTS

We are grateful for financial support from Texaco, Inc. and from Exxon Production Research Company.

Please address all inquiries to E. H. M.

REFERENCES

1. G. W. Hodgson, in Ann. N. Y. Acad. Sci., edited by A. Adler (New York Academy of Science, N. Y. 1973), Vol. 206, pp. 670–683.

2. A. Treibs, Angew. Chem., 49, 682–686 (1936).

3. Baker, E. W. J. Amer. Chem. Soc., 88, 2311–2315 (1966).

4. Baker, E. W., and Louda, J. W. Advances in Organic Geochemistry-1981, edited by M. Bjoroy (Wiley, Chester, PA, 1982) pp. 401–421.

5. Baker, E. W. and Louda, J. W. Initial Reports of the Deep Sea Drilling Project, edited by J. R. Curray et al (U. S. Government Printing Office, Washington, D. C., 1982) Vol. 64, pp. 789–814.

6. Palmer, S. E. and Baker, E. W., Science, 201, 49–51 (1978).

7. Louda, J. W. and Baker, E. W. in Initial Reports of the Deep Sea Drilling Project, edited by R. S. Yeats, et al. (U. S. Government Printing Office, Washington, D. C., 1981), Vol. 63, pp. 785–818.

8. P. P. Zelmer and E. H. Man, Org. Geochem., 5, 43-49 (1983).

9. P. P. Zelmer and E. H. Man, Org. Geochem., 7, 223-229 (1984).

10. D. D. Lavalle, E. Huggins and S. Lee, in Inorganic Reactions in Organized Media, edited by S. L. Holt, ACS Symposium Series No. 177 (American Chemical Society, Washington, D. C., 1982.)

11. P. Hambright, Coord. Chem. Rev., 6, 247-268 (1971).

12. F. R. Longo, E. M. Brown, and D. J. Quimby, in Ann. N. Y. Acad. Sci., edited by A. Adler, (New York Academy of Sciences, N. Y., 1973), Vol. 206, pp. 459-481).

13. Solvent exchange reference.

14. R. C. Weast, (editor), Handbook of Chemistry and Physics - 51st Edition (The Chemical Rubber Co., Cleveland, Ohio 1979-71) p. 232

15. S. Cady and T. J. Pinnavai, Inorg. Chem., 17, 1501-1507 (1978).

16. Stern, ref.

Appendix A

List of compounds used in Figure 3 and their abbreviations:

MPP($-CO_2$)   7-ethyl-7-despropiomesopyropheo-
           phorbide

PE($-CO_2$)   7-ethyl-7-despropiophylloerythrin

DOMPP($-CO_2$)
           7-ethyl-7-despropiodeoxymesopyro-
           pheophorbide

DPEP   deoxophylloerythroetioporphyrin

DOMPP   deoxomesopyropheophorbide-a methyl
           ester

DPE   deoxophylloerythrin methyl ester

SOME FEATURES OF PYROLYSIS-GAS CHROMATOGRAPHY-MASS
SPECTROMETRY OF KEROGENS FROM CONTINENTAL
SOURCE ROCKS IN CHINA

YUYING TONG, FONGYING JIA
Scientific Research Institute of Petroleum
Exploration and Development, Beijing
P.O. Box 910
Beijing, China.

The kerogens isolated from the continental source
rocks of oligocene in Nanyang Basin and Jurassic in
Eerduosi Basin have been analyzed. The pyrolysis
products of the kerogens are identified by pyrolysis-gas chromatography-mass spectrometry. This
study has shown that the products of the three main
types of kerogens can be characterized by pyrolysis-
GC-MS. The structural features of pyrolysed hydrocarbon molecules may be closely linked to original
organic matter in source rock. Thus, we can obtain
some useful information on the composition of the
source material and the environmental conditions
under which it was deposited. The differences in
peak shapes between pyrolysis products from three
kinds of kerogens become more obvious in the carbon
number range below C_{19}. The products of type-1
kerogen in the range below C_{19}.are dominated by a
homologous series of n-alkene/n-alkane doublets.
The products of type-III kerogen are dominated by
alkylbenzenes, phenols, naphthalenes and isoprenoid
alkenes and alkanes.

1. INTRODUCTION

As is well known, kerogen, defined as the insoluble portion
of the organic matter in sediments, is by far the most abundant form of organic carbon on Earth. Not only is kerogen the
most abundant form, but for organic geochemical purposes, it
is the most stable form of indigenous organic matter in sediments. If the structures of kerogen could be unravelled, we
would obtain much useful information on the composition of
the source material and the environmental conditions under
which it was deposited. There is a problem in obtaining
this information from kerogen, its analytical inaccessibility.
Chemical degradation has been used by some researchers. Although shortcomings exist in these methods, they can be over-
come to a great extent using pyrolysis gas chromatography-
mass spectrometry. This paper is a study of the pyrolysis
products of kerogens from continental source rocks in China.

2. SAMPLES

A brief description of the samples used in this study and
the results from a separate study on the same samples are
given in Table I. Most of the samples are abtained from
Hetaoyuan formation 2 and 3 of Oligocene age, Nanyang Basin,
Henal Province, China. Samples B-1, B-2, B-3, B-4 are
grey to dark-grey mudstones. Sample E-5 is from Yunan group
of lower Jurassic age in Eerduosi Basin, China.

3. EXPERIMENTAL

Rock samples are ground, homogenized, air-dried, and then
extracted with organic solvents (chloroform or benzene-
alcohol). The insoluble materials are treated with a mixture
of HCl/HF, then poured into an aqueous zinc bromide solution
with a density of 2.1. The separate upper layer is washed

TABLE 1 Sample descriptions and analytical data

Sample number	Location	Depth (M)	Geo. age	Type	Lithology	Kerogen				
						S_2/S_3	Cp/TOC(%)	Tmax	H/C	O/C
B-1	Nanyang Basin	1659-1661	Oligocene He-2	I	dark grey mudstone	50.28	70	433	1.45	0.15
B-2	Nanyang Basin	978-987	Oligocene He-2	II	dark grey mudstone	19.04	41	430	1.50	0.33
B-3	Nanyang Basin	1117-1117.25	Oligocene He-2	II	light-grey mudstone	9.76	30	434	1.16	0.29
B-4	Nanyang Basin	2167.57-2169.85	Oligocene He-3	II	dark grey mudstone	4.07	22	437	1.04	0.19
E-5	Eerduosi Basin	980	Jurassic yan-7	III	black mudstone	3.41	10.16	441	0.62	0.08

Note: 1. All data in Table I were provided by Wu Liyan (1986).

2. Kerogen types were determined by Rock-EVAL.

3. All samples are from core.

4. Symbols: Cp- effective organic carbon

TOC- total organic carbon

S_2/S_3- index of type

Tmax- CVEN temperature at S_2 peak

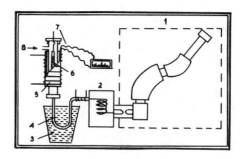

FIGURE 1   Chart of construction of the pyrolysis-GC-MS
            1- MS   2- GC
            3- Liquid nitrogen(N_2) 4- Cooling U-tube
            5- Heating element   6- Sample and sample tube
            7- Pyrometer couple   8- Carrier gas(He)

with water, extracted with chloroform, and dried. The organic residues are kerogens.

The pyrolysis system having a stainless steel reactor
is shown in Figure 1. The experiment is carried out in two
steps:

(1) preparation of kerogen pyrolysis products; (2)
introduction of pyrolysis products into the GC. About 15 mg
of kerogen are introduced into the glass sample tube (Fig.
1). Then the sample tube in hung on the hook in the heater.
At the same time, liquid nitrogen is poured into the cup.
After two minutes the heater is raised rapidly to a temper-
ature of $300^{\circ}C$, and then from $300^{\circ}C$ to $500^{\circ}C$ in ten minutes
(average rate of temperature rise: $20^{\circ}C/min$). The heater
is kept at $500^{\circ}C$ for 10 minutes. The pyrolysis products
are cooled and condensed in the U-tube. The liquid nitrogen
is then replaced by a heater with a rapid heating rate to
$350^{\circ}C$. The volatilized portions, estimated to be forty or
fifty percent of the kerogen by weight, are injected directly

into the GC-MS.

4.   RESULTS AND DISCUSSION

Figure 2 is an ion chromatogram of the pyrolysis products
called a Reconstructed Ion Chromatogram (RIC).

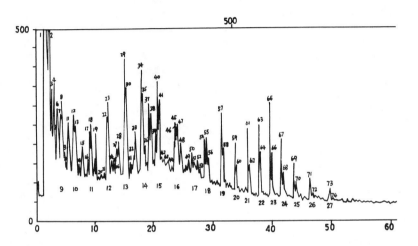

FIGURE 2 Reconstructed Ion Chromatogram (RIC) from
the pyrolysis-GC-MS of kerogen B-3, Nanyang
Basin

Note : Peak numbers refer to the compounds
listed in Table II.

The most abundant pyrolysis products of the kerogen are
n-alkanes, n-alk-l-enes, aromatics, alkylphenols, isoprenoid
alkenes and alkanes.   N-alkanes and n-alk-l-enes are identified using mass chromatogram series m/e 57, 71, 55, 69, . . .
and 240, 268, 266, . . . , together with their corresponding
mass spectra.   Identification of pristene and pristane is on
the basis of their molecular ions m/e 266 and 268.   Similarly alkyl-benzenes can be identified using m/e 91, 92, 106,

TABLE II   Pyrolysis products of kerogen (B-3)

Peak No.	Compounds Identity	M.W.	Peak No.	Compounds Identity	M.W.
1	carbon dioxide and hydrocarbons(C_3–C_4)		38	octylbenzene	190
2	hydrocarbons (C_4–C_6)		39	butylindole	173
3	n–octane	114	40	n–pentadecane	212
4	oct–1–ene	112	41	pentadec–1–ene	210
5	benzene	78	42	dimethylnaphthalene	156
6	toluene	92	43	unknown	
7	n–nonane	128	44	propylnaphthalene	170
8	non–1–ene	126	45	n–hexadecane	226
9	ethylbenzene	106	46	hexadec–1–ene	224
10	unknown		47	methylethylnaphthalene	170
11	propylbenzene	120	48	C_9–benzene	204
12	n–decane	142	49	isoprenoid C_{18}–ALKANE	254
13	dec–1–ene	140	50	n–heptadecane	240
14	phenol	94	51	heptadec–1–ene	238
15	tetramethylbenzene	134	52	prist–1–ene	266
16	propylmethylbenzene	134	53	prist–2–ene	266
17	n–undecane	156	54	phytane	282
18	undec–1–ene	154	55	n–octadecane	254
19	ethylpropylbenzene	148	56	octadec–1–ene	252
20	cresol	108	57	n–nonadecane	268
21	naphthalene	128	58	nonadec–1–ene	266
22	n–dodecane	170	59	n–eicosane	282
23	dodec–1–ene	168	60	eicos–1–ene	280
24	triethylbenzene	162	61	n–heneicosane	296
25	propylphenol	136	62	heneicos–1–ene	294
26	methylnaphthalene	142	63	n–docosane	310
27	isoprenoid C_{14}–alkane	198	64	docos–1–ene	308
28	diethylpropylbenzene	176	65	n–tricosane	324
29	n–tridecane	184	66	tricos–1–ene	322
30	tridec–1–ene	187	67	n–tetracosane	338
31	butylphenol	150	68	tetracos–1–ene	336
32	heptylbenzene	176	69	n–pentacosane	352
33	isoprenoid C_{15}alkane	212	70	pentacos–1–ene	350
34	n–tetradecane	198	71	n–hexacosane	366
35	tetradec–1–ene	196	72	hexacos–1–ene	364
36	isoprenoid C_{15}–alkene	210	73	n–heptacosane	380
37	isoprenoid C_{16}–alkane	226	74	heptacos–1–ene	378

Note:

    1.^{GC} conditions: Capillary column OV–17 25m x 0.23mm

                  temperature 50 – 280 °C at 4 °C / min.

    2.   FIGURES 3 and 5 show some of the corresponding mass spectra

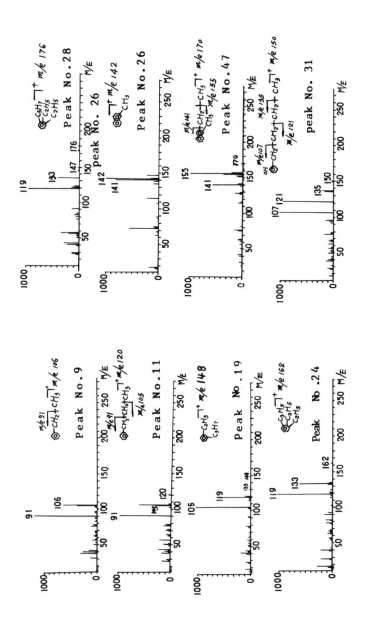

FIGURE 3   Mass spectra of kerogen products

120, . . . , and m/e 105, 106, 119, 120, 133, 134, . . . mass chromatograms. m/e 107, 108, 121, 122, 135, 136, . . . are characteristic of alkylphenols, and the peaks at m/e 141, 142, 155, 156, 169, 170 are of alkylnaphthalenes. Consequently, structural features and molecular weight are both determined. The identifications and configurations of the kerogen pyrolysis products are shown in Figure 2 and Table II.

A comparison of pyrolysis products from three kinds of kerogens is informative. Figure 4 is a RIC diagram of pyrolysis products from representatives of the three kerogen types I, II and III. All three types are characterized by a bimodal distribution of two humps (refer to dotted line in Fig. 4). The pyrolysed hydrocarbons in the carbon number range below C_7 to C_{19} constitutes the first hump, while the distribution of the peak group after C_{19} forms the second hump. Generally, the pyrolysis products are mainly n-alkanes and n-alk-1-enes. Both of them are present in the products from the three kinds of kerogens. The differences between pyrolysis products from the three kinds of kerogens becomes more obvious in the distribution of components in the carbon number range below C_{19}.

Figure 4 shows that the products in the carbon number range below C_{19} of type-1 kerogen (B-1) are dominated by a homologous series of n-alkene/n-alkane doublets which extends to about C_{28} (about 77.9%). The other pyrolysis products, such as aromatics and alkylphenols (about 20.3%), though definitely present, are only moderately important in relation to the doublet homology. The pyrolysis products in the carbon number range below C_{19} of the type-III kerogens (E-5) show the typical characteristics of the kerogens originating from higher plants. They are donimated by alkyl-benzenes, phenols,

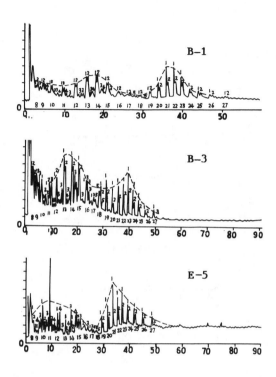

FIGURE 4 Ion chromatogram of the pyrolysis–GC–MS runs
from three types of kerogens in the continental
source rock

1– alkanes   2– n–alk–1–enes

3– aromatics   4– alkylpheols

5– isoprenoid alkenes
and alkanes

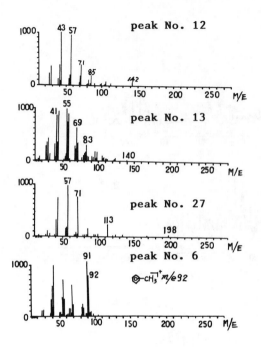

FIGURE 5   Mass spectrum of kerogen products

naphthalenes, isoprenoid alkenes and alkanes (about 50.7%).
The homonogous series of n-alkane/n-alkene doublets are
relatively less abundant (accounting for only about 48.2%).
Pyrolysis products of type-II kerogens (B-3) show charac-
teristics intermediate between type-I and type-III (Table
III).

Table III gives characteristics of the pyrolysis products from different types of kerogens.

Parameter P-1 shows that the short chain hydrocarbons
dominate in the pyrolysis products of type-I kerogen, but
the longer chain hydrocarbons dominate in the pyrolysis products of type-III kerogen (E-5).   Parameter P-1 for the py-

TABLE III   Comparison of pyrolysis products
from three types of kerogen in continental
source rocks

sample	location	type of kerogen	P–1	P–2	P–3	P–4	P–5	P–6
B–1	Nan Yang	I	77.9	20.3	13.6	6.7	56.4	39
B–3	Nan Yang	II	74.9	23.5	20.3	3.2	49.1	32
B–5	Eerduosi	III	48.2	50.7	25.9	24.8	30.9	47.3

Note : 1. The types are determined according to ROCK–
EVAL.

2. All data are determined by peaks height in
RIC.

P–1 : The ratio of Σ n–alkanes and n–alk–1–enes to
Σ products in the carbon number range below C_{19} .

P–2 : The ratio of Σ aromatic+alkylphenols+isoprenoid
alkenes and alkanes to Σ products in the carbon
number range below C_{19} .

P–3 : The ratio of Σ aromatics to Σ products in the
carbon number range below C_{19} .

P–4 : The ratio of   alkylphenols+isoprenoid alkenes
and alkanes to Σ products in the carbon number
range below C_{19} .

P–5 : The ratio of Σ n–alkanes+n–alk–1–enes in the
carbon number range below C_{19} to Σ all pyrolysis products.

P–6 : The ratio of Σ n–alkanes+n–alk–1–enes in the
carbon number range above C_{19} to Σ all pyrolysis products.

rolysis products of type-II kerogen lies between I and III.

From parameter P-4 in Table III we can see clearly
that the alkylphenol contents in the pyrolysis products of
the type-III kerogen (E-5) is four to seven times that of
the type-I (B-1) and type-II (B-3). The abundant alkylphenols suggest that sample E-5 is taken from shallow lake
marshland coal-bearing continental sediments in the Eerduosi
Basin. The alkylphenols come from coal-bearing strata whose
source materials are from terrestrial higher plants. Sample
B-1 is obtained from the deeper lake sediments in Nanyan
Basin, in which there were more freshwater algae, pediastrum,
and typical fossils of freshwater carp fish. The kerogen
originated mainly from various types of algae and aquatic
plants which are abundant in aliphatic structures and could
produce more aliphatic hydrocarbons. Sample B-3 is taken from
the lake sediments in Nanyan Basin. Some of the organic matter in the sediments were provided by aquatic organisms and
algae. In addition, the remnants of terrestrial plants are
also evident. The parameters for B-3 (type-II) in Table III
mostly appear between the other two, but are more closely
aligned with those of sample B-1 (type-I), possibly due to
a greater contribution to the pyrolysis products from algal
remains than from those of terrestrial plants.

5. CONCLUSIONS

The pyrolysis compounds from different types of kerogen in
the carbon number range below C_{19} possess the following
characteristics: type-I kerogen derived from aquatic organisms yields mainly aliphatic hydrocarbons with relatively
short n-alkane and n-alkene carbon chains; type-III kerogen,
derived from terrestrial higher plants produces aromatics,
alkylphenols and isoprenoid alkenes and alkanes in signifi-

cantly higher amounts than other kerogens; type-II kerogen, derived from a mixture of the above two source materials, shows characteristics between type-I and type-III, but closer to type-I.

PRESENCE OF BENZOHOPANES, MONOAROMATIC SECOHOPANES, AND SATURATE HEXACYCLIC HYDROCARBONS IN PETROLEUMS FROM CARBONATE ENVIRONMENTS

G. G. L. RINALDI, V. M. LEOPOLD, AND C. B. KOONS
Exxon Production Research Company
P. O. Box 2189
Houston, TX   77252-2189

Different homologous series of biomarkers related to the hopane carbon skeleton have been observed in the monoaromatic fraction of petroleums of Paleozoic age associated with carbonate environments. A hexacyclic homologous series of benzohopanes (C_{32} -C_{35}) has been found to be particularly abundant in this type of geological environment. This series, which has been previously characterized by Hussler [1,2], et.al. (1984, 1985) and Belayouni[3], et.al. (1985) originates during very early diagenesis from C_{35} bacteriohopanetetrol precursors. On the other hand, a homologous series of monoaromatic tetracyclic secohopanes are generated by maturation. Their formation requires the opening of ring C from the original hopane precursors, by breaking the C_8 --C_{14} bond and subsequent aromatization of the D ring. Finally, a novel homologous series of saturate hexacyclic hydrocarbons (C_{31} - C_{35}) which are also derived from C_{35} hopanoid precursors is found in these petroleums. A generalized summary about the geochemical significance of these different homologous series is given.

1. INTRODUCTION

It has already been shown (Seifert and Moldowan[4], 1978)

that biological marker compounds of the sterane and triterpane type, can be successfully applied in oil-oil and oil-source correlation studies because they are particularly powerful as source and maturity indicators. Up to date, one of the most interesting aspects of research in this particular field is the understanding of the geochemical mechanisms involved in the aromatization of sterane and triterpane hydrocarbons in the various stages of the diagenetic process.

Several researchers have shown (Meinschein[5], 1959; Orr[6], et.al., 1967; Aizenshtat[7], 1973; Greiner[8], et.al., 1976; Spyckerelle[9,10], et.al., 1977 a,b; LaFlamme[11], et.al. 1978; Wakeham[12], et.al., 1980) that aromatic hydrocarbons can be formed during early diagenesis as extra-cellular by-products of anaerobic bacterial metabolism which uses isoprenoid precursors as organic substrates. From the same precursors, as diagenesis progresses, abiotic reactions form aromatic hydrocarbons. Basically, the reaction mechanisms are those of disproportionation and dehydrogenation. From an energetic viewpoint, disproportionation and partial fragmentation of terpenoids are the preferred abiotic reactions because they are highly exothermic. From an analysis of the monoaromatic fraction of oils from the Tertiary of the

Gulf Coast, Bendoraitis[13] (1974) proposed a convincing scheme for the sequential aromatization and cleavage of pentacyclic triterpenoids into aromatic secotriterpanes. Further cleavage of the latter compounds would generate some of tetrahydronaphthalene and indane hydrocarbons found in petroleums. Seifert[14], et.al. (1983) showed in the laboratory that monoaromatic steroid hydrocarbons are formed by dehydrogenation of sterane hydrocarbons precursors.

MacKenzie[15], et.al. (1981) in a study that was carried out with rock samples from the Paris Basin found that with increasing maturity monoaromatic steroids were transformed (by dehydrogenation reactions) into tri-aromatic steroid hydrocarbons. The samples were selected from a few wells at different depth intervals.

It is the purpose of this paper to show the occurrence and geochemical significance of two series of monoaromatic biomarkers which can be easily related to hopanoid precursors.

One series is made up of benzohopane homologs and the other one is constituted of monoaromatic secohopane hydrocarbons. The former series is formed during the early diagenetic process by anaerobic bacterial action, whereas the latter one is formed by maturation. The benzohopane series has been previously identified and characterized by Hussler[1,2], et.al. (1984, 1985) and Belayouni[3], et.al. (1985). The monoaromatic secohopane homologous series has also been previously characterized (Hussler[2], et.al. 1985).

Finally, a novel hexacyclic saturate hydrocarbon (containing thirty-one carbons) has been tentatively identified. It is the first and also the most abundant member of a homologous series of hexacyclic hydrocarbons (C_{31} - C_{35}). There is strong evidence that this homologous series is probably formed under anaerobic conditions during early diagenesis from bacteriohopanetetrol precursors.

2. EXPERIMENTAL

The oils employed in this study (five) are from Paleozoic carbonate reservoirs of Eastern Montana.

After deasphaltening, the pentane soluble portion of the oils was separated by High Pressure Liquid Chromatograph (Waters Associates) into saturates and aromatics. The column used was silica bonded with amino groups (Waters Associates) as the stationary phase and n-hexane as the mobile phase.

The saturate fraction was subjected to urea adduction to isolate the branched/cyclic "cut" for GC/MS analysis.

The procedure for further separating the aromatic fraction was carried out by using a HPLC procedure originally used by Radke[16], et.al. (1983). The whole aromatic fraction was separated according to number of aromatic rings into mono-, di-, tri-, and tetra- aromatics as shown in fig. 1.

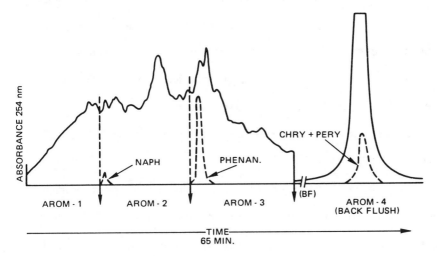

FIGURE 1.   HPLC separation by ring number of the aromatic fraction.

The monoaromatic fraction was dissolved in methylene chloride and injected "on column" onto a Carlo Erba FTV 2900 gas chromatograph fitted with a fused silica capillary column (60m X.3mm i.d.) coated with SE-54 (1μ).

Helium was used as carrier gas. Typical temperature programming conditions were 50 - 320° C at 3° C/min. The column effluent was split (1:1) and detected both by FID (Flame Ionization Detector) and FPD (Flame Photometric Detector). The latter detector is specific for sulfur containing hydrocarbons.

GC/MS analyses of the monoaromatic fraction were carried out in the electron impact mode (EI). A Carlo Erba 2900 FVT equipped with an "on column" injector was interfaced directly to an Extranuclear quadropole mass spectrometer. The gas chromatograph was fitted with a fused silica capillary column (60m X.3mm i.d.) coated with Se-54 (1μ). Helium was used as carrier gas. Again, temperature programming was typically 50 - 320° C at 3° C/min.

The operating conditions of the mass spectrometer were the following: electron energy, 70 ev., filament current, 30 μ A, accelerating voltage, about 2KV, and source temperature 200° C.

Data acquisition (scan time 1 sec) and processing, were carried out with a TEKNIVANT data system.

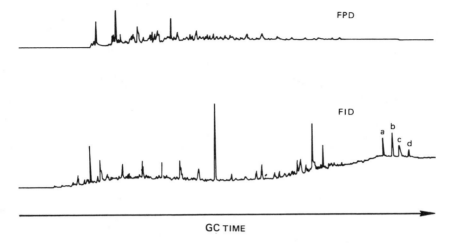

FIGURE 2.   Double trace gas chromatogram (FID, FPD) for
            the monoaromatic fraction. The labeled peaks
            in the FID trace indicate the homologs

3. RESULTS AND DISCUSSION

3.1 Benzohopanes

Fig. 2 shows the capillary GC traces (FID & FPD) of the
monoaromatic fraction.   The peaks labeled a, b, c and d
correspond to benzohopane homologs, whose structures are
shown in fig. 3.   This homologous series ranges from C_{32}
to C_{35}.   Figs. 4 and 5 show the mass spectra of the C_{32}
and C_{33} homologs, respectively.   As can be seen, m/z 191
is the base peak, as in the regular   17 a (H) ⎮ hopane
series.   The other major fragment ions m/z 211 + 14n and
226 + 14n (n = 0-3) indicate the presence of an aromatic
ring in the right hand portion of the benzohopane

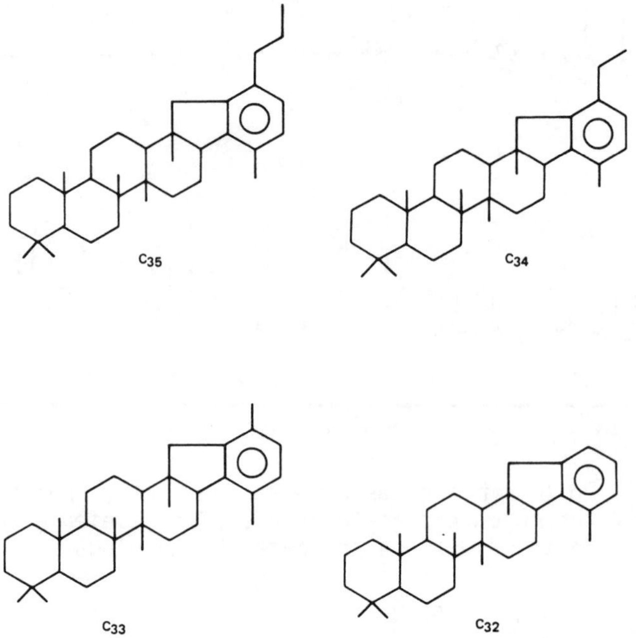

FIGURE 3.   Structures of the benzohopane homologous series (C_{32} - C_{35}).

molecule (Hussler[1,2], et.al. 1984, 1985). The spectrum also contains other intense fragment ions in the neighborhood of m/e 144 + 14n and 156 + 14n which are attributable to substituted indane fragment ions ejected from the hexacyclic structure under electron impact. These latter fragment ions do suggest that a benzene ring is condensed to the E ring of a hopane skeleton. Hussler[1,2], et.al. (1984, 1985) have established that the benzohopanes are hydrocarbon derivatives of hopanoid precursors, such as

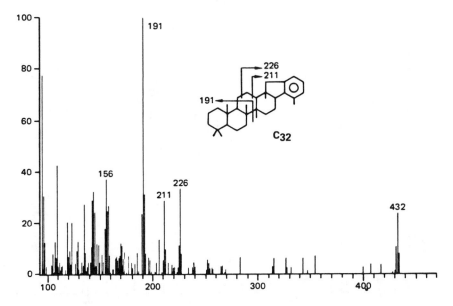

FIGURE 4.   Mass spectrum of the C_{32} benzohopane homolog.

the polyalcohol bacteriohopanetetrol. This precursor is an important constituent of the cell membrane of many procaryotic organisms (Ourisson[16], et.al. 1979). They postulate a series of dehydration and cyclization reactions taking place on the bacteriohopanetetrol precursor (Fig. 6) during very early diagenesis to form the C_{35} benzohopane hydrocarbon. However, they envision inorganic processes (acid catalysis mediated by clays) for the subsequent degradation of the side chain of the C_{35} benzohopane hydrocarbon to form the lower benzohopane homologs. We agree with the authors' model for the

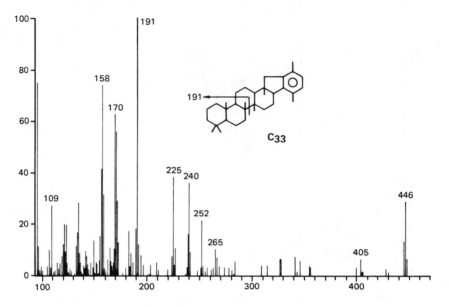

FIGURE 5.   Mass spectrum of the C_{33} benzohopane homolog.

formation of the C_{35} hydrocarbon.   However, for the formation of the lower benzohopane homologs we propose an alternative hypothesis.   In geochemical studies of carbonate environments carried out at Exxon Production Research Co., we did not observe any significant relationship between type or amount of clay and the abundance of different benzohopane homologs relative to each other. Thus, we prefer to formulate the hypothesis that the side chain shortening is generated by biological oxidation prior to the dehydration reaction, leading to benzohopane

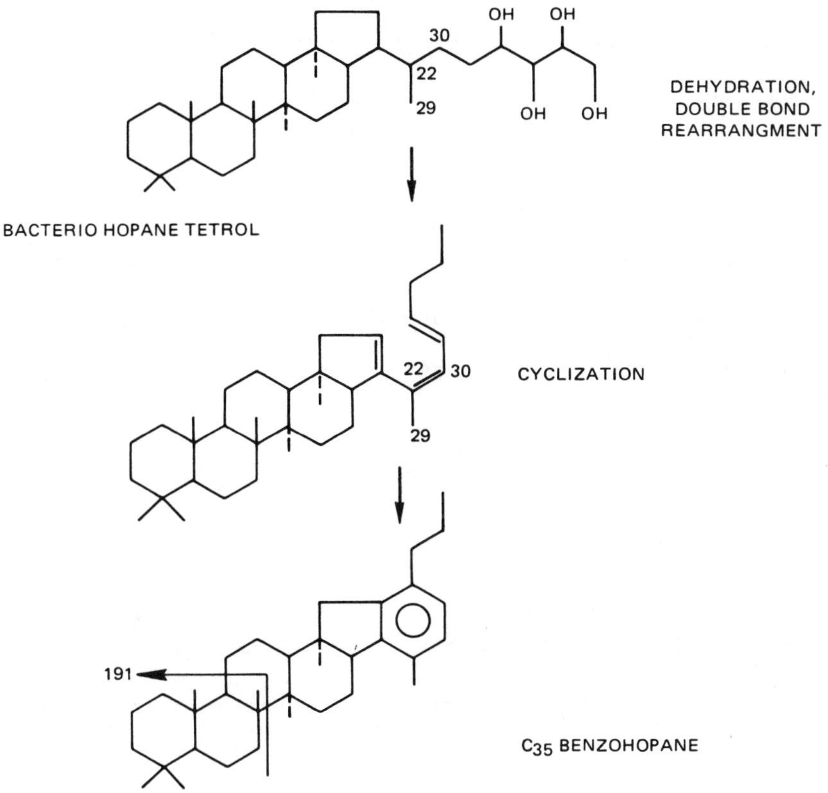

FIGURE 6.   Postulated early diagenetic transformation
of bacteriohopanetetrol to C_{35} benzohopane
hydrocarbon.

hydrocarbons. The last step (aromatization) is carried out by bacterial oxidation under anaerobic conditions. Thus, this class of monoaromatic biomarker is not maturation induced but is formed during very early diagenesis by bacterial action.

3.2 Ring D Monoaromatic 8, 14 Secohopanes

The peaks labeled with C_{27}, C_{29}, C_{30} in fig. 7 are members of the major series of monoaromatic secohopanes (C_{27} -C_{35}) found in petroleums which has already been characterized. (Hussler[2], et.al. 1985).

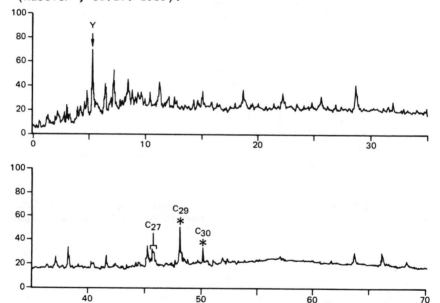

FIGURE 7. TIC trace of the monoaromatic fraction, showing the monoaromatic 8, 14, secohopanes (C_{27}, C_{29}, C_{30}) and a possible tetra-alkylated indane (V).

Fig. 8 shows the mass spectrum of the most abundant homolog as an example of this series. It contains twenty-nine carbon atoms (M^+ = 394). The main features of that series is a major m/z = 365 (base peak), but absence of m/z = 123 fragment ion. Schmitter [18], et.al. (1982) have shown that m/z = 123 (loss of ring A) is the base peak for the 8,14 secohopane hydrocarbons. We do not have

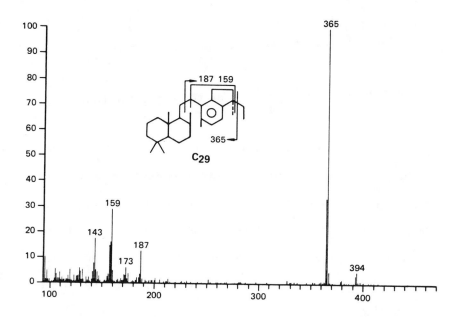

FIGURE 8.   Mass spectrum of the C_{29} monoaromatic 8, 14 secohopane.

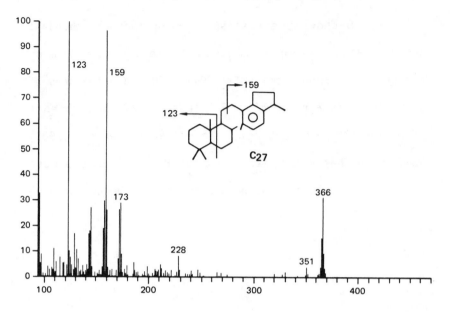

FIGURE 9.   Mass spectrum of the C_{27} monoaromatic 8, 14 secohopane.

any reasonable explanation for the absence of this important fragment ion from the mass spectra of the C_{29}, C_{30}, C_{31}, etc. members of this major monoaromatic seco-hopane series. Fig. 9 shows the mass spectrum for the C_{27} homolog (M^+ = 366). As can be seen it has a large m/z = 123 fragment ion which should be expected from the loss of the A ring.

As far as the C_{29} member is concerned, the m/z = 365 fragment probably represents the facile loss of the side chain from the C_{21} carbon, whereas, the fragment ions m/z = 145, 159 and 173 represent mono-, di- and tri-alkylated indane fragments (Bendoraitis[13], 1974; Hussler[2] et.al. 1985). Other homologous series of 8, 14 ring D, and possibly ring A (or B) monoaromatic secohopanes seem to be present in trace amounts. However, we did not synthesize any standard to prove the above statement. Thus, our conclusions are based only on mass spectral data and general geochemical considerations.

The geochemical significance of the different monoaromatic 8, 14 secohopane homologous series is that all of them are formed as a consequence of thermal alteration. In addition, they can fragment further under maturation giving rise to lower molecular weight monoaromatic hydrocarbons. Bendoraitis[13] (1974) postulated that some of the tetrahydronaphthalenes and indane isomers found in petroleums originated as cleavage products of terpenoids. Hussler[2], et.al. (1985) reproposed this same concept by stating that the monoaromatic secohopanes are indeed the logical precursors of some of these low molecular weight monoaromatics. We are in agreement with the conclusions of the above authors. To further substantiate the above statements, we are showing the mass spectrum (Fig. 10) of a low molecular weight monoaromatic

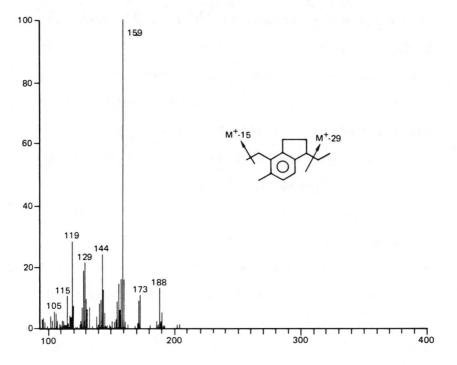

FIGURE 10.   Mass spectrum of a probable C_{14} tetra-alkyl indane.

hydrocarbon. The peak labeled Y (Fig. 7) is the one from which the mass spectrum was obtained. We assume this abundant compound to have an indane structure. This assumption would be in line with the idea that the homologous series of ring D monoaromatic secohopanes contains the most important precursors of lower molecular weight monoaromatics having an indane carbon skeleton. Based on previous work (Bendoraitis[13], 1974; Hussler[1,2], et.al., 1984, 1985), we formulate a generalized model for

the formation of lower molecular weight monoaromatics
(Fig. 11). Clearly, maturation seems to be the most
important geochemical process which generates some of
these compounds from high molecular weight triterpenoid
hydrocarbons. However, further systematic work is
required to identify many of these lower molecular weight
structures in order to establish precursor-product
relationships. This is particularly important for the
understanding of their potential geochemical significance
in condensate-condensate correlation work.

3.3 18 a (H), Hexacyclic Hydrocarbon

An investigation of the saturate fraction (urea non-
adduct) from this particular group of oils indicates that
the sterane class is present in a trace amount, whereas
the pentacyclic triterpane class was practically the sole
contributor to this fraction.

Fig. 12 shows the TIC of the urea non-adduct. As can
be seen, the C_{29} norhopane is the most abundant compound
followed by the C_{30} hopane. The peak marked with an
asterisk corresponds to a novel hexacyclic compound
containing thirty-one carbon atoms. This compound is the

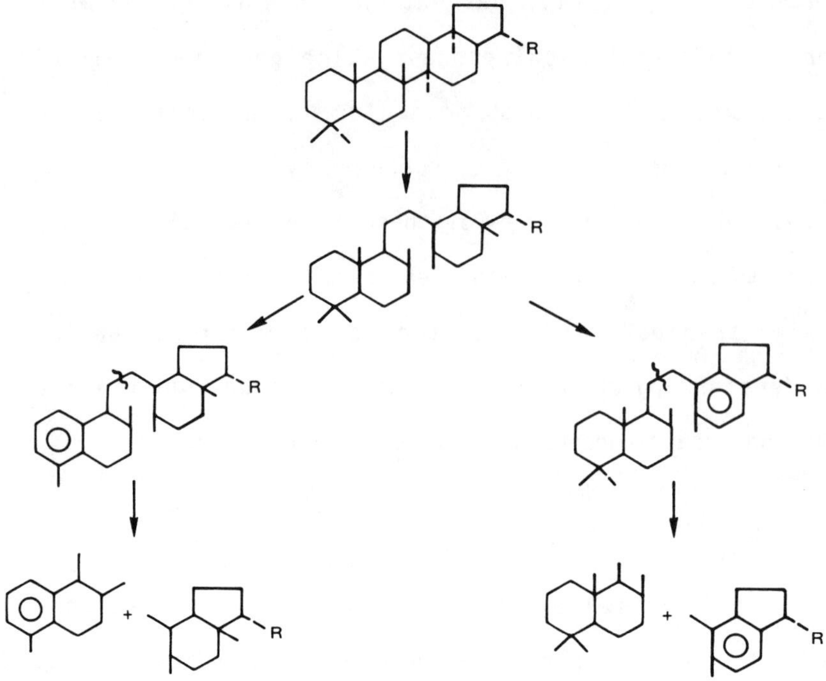

FIGURE 11.   Possible route of formation of the monoaromatic secohopanes and low molecular
weight monoaromatics from hopanoid
precursors.

first (and most abundant) member of a homologous series of
hexacyclics containing up to thirty-five carbon atoms.
Fig. 13 shows the mass spectrum of this novel compound.

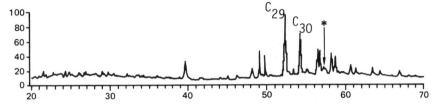

FIGURE 12.   TIC of the saturate fraction (urea-non-
              adduct).

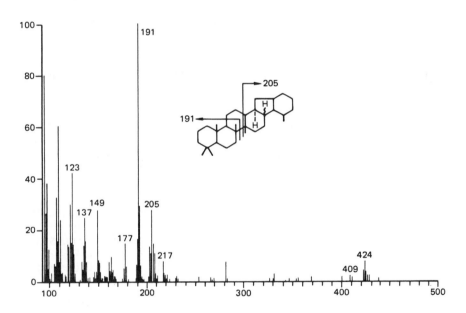

FIGURE 13.   Mass spectrum of a novel hexacyclic C_{31}
              hydrocarbon.

A hopanoid structure is indicated by m/z = 191 as the base peak. The other significant fragment ion which can give clues about the potential hopanoid precursors is m/z = 217.

Moldowan, [18] et.al. (1984) have shown by x-ray crystallography of isolated pure crystalline material that this fragment ion is related to the 18 a (H), 28,30-bisnor-hopanes found in Monterey oils.

In light of that piece of information we would like to formulate its structure based upon 17a (H), 18a (H) sterochemistry. As already said, no steranes are present in this sample. Thus, the possibility of the m/z = 217 as a "contaminant" fragment ion from those latter compounds is not a problem.

As far as the origin of this class of compounds is concerned, the molecular size of these hydrocarbons is too large to be bacterial cell membrane constituents. Like the benzohopanes they are formed from bacteriohopanetetrol (C_{35}) precursors during the very early diagenetic process.

One question which comes to mind is the following: What controls the total abundance of the benzohopanes which are also a product of early diagenesis, relative to the saturate hexacyclic hydrocarbon class? We propose that after biological oxidation of the side chain and the dehydration and cyclization reactions have occurred, the hypothetical unsaturated hexacyclic intermediate can be preferentially converted into benzohopanes if the physical chemical environment of deposition is very reducing (low Eh) and the pH of the pore water is highly alkaline (9.0). On the other hand, if the environment is still very reducing but more hydrogen rich (pH = 8.0) the unsaturated intermediate is preferentially converted into saturated hexacyclic hydrocarbons. The former situation would be characteristic of mesosaline lagoons where cyanobacterial systems thrive, whereas the latter environment would be characteristic of many prodelta sediments in which organic matter was deposited near or at average salinity conditions. This particular set of oils was selected from carbonate environments of the former type. This is probably the reason why the monoaromatic benzohopanes are so abundant relative to the hexacyclic saturate hydrocarbons. Additional research is in progress to further substantiate the above hypothesis.

4. CONCLUSIONS

1. The benzohopanes are formed during early diagenesis from a bacteriohopanetetrol precursor.

2.   Maturation induces the formation of aromatic secohopanes from hopanoid precursors.

3.   Some of the monoaromatic bicyclics are formed by maturation from aromatic secohopanes.

4.   A novel hexacyclic compound with a hopane carbon skeleton has been isolated. Probably it is formed during early diagenesis from cyclization of the side chain of extended hopanoids.

5.   ACKNOWLEDGEMENTS

We would like to thank the Management of Exxon Production Research Company for their support and kind permission to publish this paper.

REFERENCES

1.   G. HUSSLER, P. ALBRECHT, G. OURISSON, M. CESARIO, J. GUILHEM, AND C. PASCARD, Tetrahedron Letters, 25 (11), 1179 - 1182, (1984).
2.   G. HUSSLER, J. CONNAN, AND P. ALBRECHT, Novel Families of tetra and hexacyclic aromatic hopanoids predominant in carbonate rocks and crude oils, Advances in Organic Geochemistry 1983 edited by P. A. Schenck, J. W. DeLeew, and G. W. M. Lijmbach (Pergamon Press, Oxford), 39 - 49, (1985).
3.   H. BELAYOUNI AND J. TRICHET, Hydrocarbons in phosphatized and nonphosphatized sediments from the phosphate basin of Gafsa, Advances in Organic Geochemistry 1983 edited by P.A. Schenck, J. W. DeLeew, and G. W. M. Lijmbach (Pergamon Press, Oxford), 741 - 754, (1985).
4.   W. K. SEIFERT AND J. M. MOLDOWAN, Geochim. et Cosmochimica Acta, 42, 77-95 (1978).

5. W. G. MEINSCHEIN, Am. Assoc. Petroleum Geologists Bull., 43, 925 - 943, (1959).

6. W. L. ORR AND J. R. GRADY, Geochim. et Cosmochimica Acta, 31, 1201 - 1209, (1967).

7. A. AIZENSHTAT, Geochim. et Cosmochimica Acta, 27, 559 - 567, (1973).

8. A. C. GREINER, C. SPYCKERELLE, AND P. ALBRECHT, Tetrahedron, 32, 257 - 260, (1976).

9. C. SPYCKERELLE, A. C. GREINER, P. ALBRECHT, AND G. OURISSON, Jour. Chem. Research, M, 3801 - 3828; S, 332 - 333, (1977b).

10. C. SPYCKERELLE, A. C. GREINER, P. ALBRECHT, AND G. OURISSON, Jour. Chem. Research, M, 3801 - 3828; S, 332 - 333, (1977b).

11. R. E. LAFLAMME AND R. A. HITES, Geochim. et Cosmochimica Acta 42, 289 - 0303, (1978).

12. S. G. WAKEHAM, C. SHAFFNER, AND G. GIGER, Geochim. et Cosmochim. Acta, 44, 415 - 429, (1980).

13. J. G. BENDORAITIS, Hydrocarbons of biogenic origin in petroleum. Aromatic Triterpenes and Bicyclic Sequiterpenes, Advances in Organic Geochemistry 1973, edited by B. Tissot, and F. Biener, (Editions Technilp, 27, Rue Ginoux, 75737 Paris), 209 - 224, (1974).

14. W. K. SEIFERT, R. M. K. CARLSON, AND J. M. MOLDOWAN, Geomimetic Synthesis, Structure Assignment, and Geochemical Correlation Application of Monoaromatized Petroleum Steroids, Advances in Organic Geochemistry 1981, edited by M. Bjorøy et.al. (John Wiley & Sons Limited, New York), 710 - 724, (1983).

15. A. S. MACKENZIE, C. F. HOFFMANN, AND J. R. MAXWELL, Geochim. Cosmochimica Acta, 45, 1345 - 1355, (1981).

16. M. RADKE AND D. H. WELTE,, The methylphenanthrene index (MPI): a maturity parameter based on aromatic hydrocarbons, Advances in Organic Geochemistry 1981, edited by M. Bjorøy et.al., (John Wiley & Sons, New York), 504 - 512, (1983).

17. G. OURISSON, P. ALBRECHT, AND M. ROHMER, Pure Appl. Chem. 51, 709 - 729, (1979).

18. J. M. SCHMITTER, W. SUCROW, AND P. J. ARPINO, Geochim. et Cosmochim. Acta, 46, 2345 - 2350, (1982).

19. J. M. MOLDOWAN, W. K. SEIFERT, E. ARNOLD, AND J. CLARDY, Geochim. et. Cosmochimica Acta, 48, 1651 - 1661, (1984).

TANDEM MASS SPECTROMETRIC ANALYSES OF GEOPORPHYRINS

J.M.E. QUIRKE and M. PEREZ
Department of Chemistry,
Florida International University,
Tamiami Trail,
Miami FL33199

E.D. BRITTON and R.A. YOST
Department of Chemistry,
University of Florida,
Gainesville, FL32611

Analyses of high carbon number geoporphyrins from Boscan oil (Cretaceous W. Venezuela) indicate that there are many isomers present, and that there are at least two sites of extended ($>C_2$) alkylation on the porphyrin macrocycle. There is tentative evidence that DPEP porphyrins with large isocyclic rings are abundant. Evidence for a novel series of benzo porphyrins without ethyl groups is also presented. MS/MS studies on the porphyrins from Gafsa Basin chert (Paleocene, Tunisia) indicates the presence of benzo DPEP porphyrins in their metal free state.

1. INTRODUCTION

The geoporphyrins occur in a wide range of geological environments as mixtures of up to at least seven skeletal types: the aetio (1), DPEP-5 (2), DPEP-6 (3), DPEP-7 (4), benzo-aetio (5), benzo-DPEP (6), and di-DPEP (7) components.

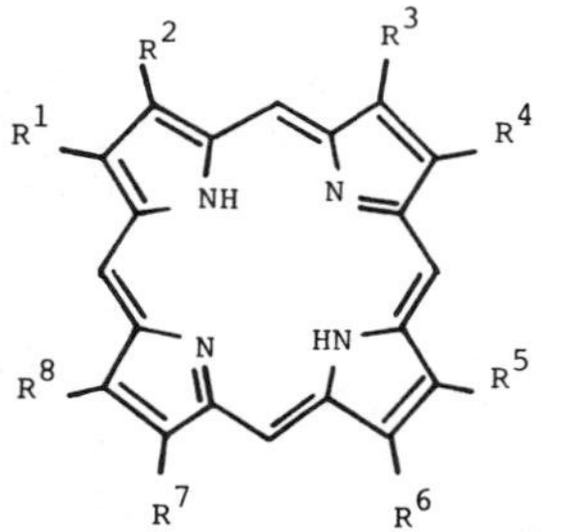

(5)

8 R = CH_3
9 R = CHO

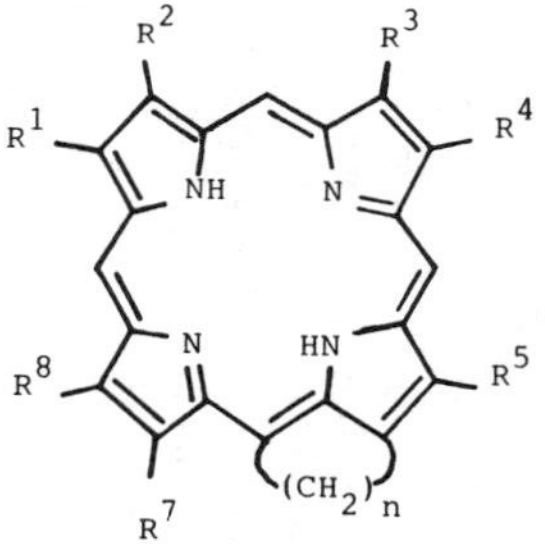

2(a) $R^1, R^3, R^5, R^8 = CH_3$; $R^2, R^4, R^7 = C_2H_5$; n = 2
(b) $R^1, R^5, R^8 = CH_3$; $R^3 = H$; $R^2, R^4, R^7 = C_2H_5$; n = 2
3 n = 3
4 n = 4

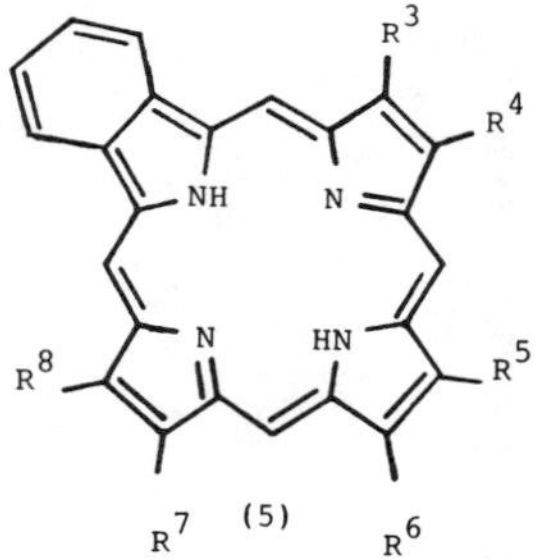

(7)

1(a) $R^1, R^3, R^5, R^8 = CH_3$; $R^2, R^4, R^6, R^7 = C_2H_5$
(b) $R^1, R^3, R^5, R^7 = CH_3$; $R^2, R^4, R^6, R^8 = n-C_5H_{11}$
(c) $R^1, R^3, R^5, R^7 = CH_3$; $R^2, R^4, R^6, R^8 = n-C_7H_{15}$

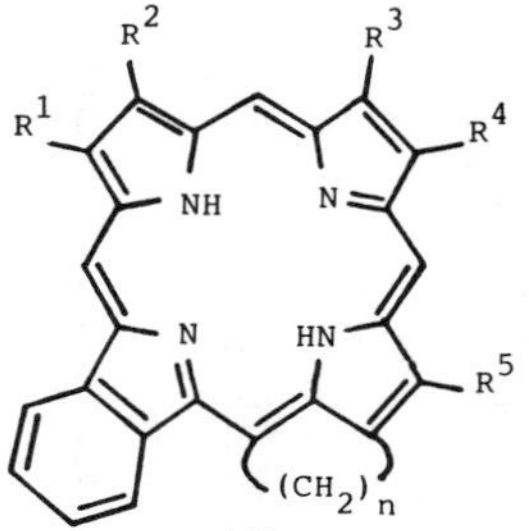

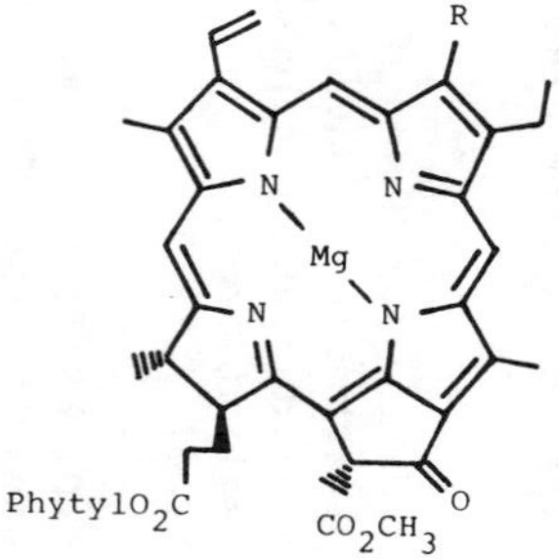

(6)

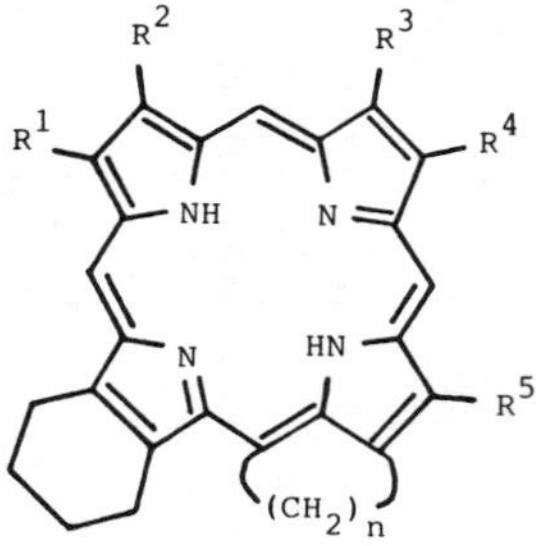

They exist primarily as nickel or vanadyl components in crude oils, oil shales and bitumens[1], but copper complexes and metal free species have also been detected.[2,3]

During the past seven years, there have been great strides made in the structure determination of the geoporphyrins in shales, and bitumens using either nuclear magnetic resonance spectroscopy (NMR) on the porphyrins themselves[4-7], or a combination of NMR and chemical ionisation mass spectrometry (CIMS)[8], or by NMR analyses of porphyrin derivatives.[9,10] In general, these studies gave support for the Treibs hypothesis[11] that the porphyrins were formed by well-defined degradative reactions of the functional groups of naturally-occurring chlorophylls. The isocyclic ring and the vinylic moiety of chlorphyll a (8) are the most reactive positions in the most abundant of all the chlorins. Nevertheless there are still many questions to be addressed e.g. the origins of the DPEP porphyrins with extended ($>C_5$) isocyclic rings, and the benzo porphyrins.

The above studies have centered on porphyrins from oil shales; there has been much less information on the nature of the petroporphyrins- porphyrins occurring in petroleum- primarily because of the complexity of the mixtures.[1,12] One of the most valuable methods employed in the analyses of the petroporphyrin mixtures is the oxidative degradation of the porphyrins to maleimides (pyrrole-2,5-diones). Such studies have revealed the presence of porphyrins with extended alkyl substituents ($>C_2$) in samples as diverse as Boscan oil (Venezuela), La Paz oil (Venezuela), Lloydminster oil (Canada) and Coalport tar tunnel (Shropshire, England).[13-16] However, the method does not provide data on individual porphyrin species. Similarly, there has also been very little data reported on the nature of the metal-free porphyrin distributions.

Tandem mass spectrometric analyses provide a technique which overcomes the difficulty of separating intractable components, and has been successfully used for the analysis of mixtures from many sources including those of the pharmaceutical, industrial, environmental, agricultural and petrochemical fields.[17] We present the results of the first tandem mass spectrometric (MS/MS) analyses on high carbon number ($>C_{33}$) geoporphyrins from Boscan oil (Cretaceous, W. Venezuela) along with compelling evidence for a novel geoporphyrin series. In addition, we outline results of analyses of metal free geoporphyrins from the Gafsa Basin of Tunisia.

2. EXPERIMENTAL

2.1 Samples

The Boscan oil was a gift from the Corporacion de Petroleos de Venezuela. The Boscan field is situated West of Lake Maracaibo, and it is believed to be formed from the La Luna formation (Cretaceous).[18,19] The geology of the Bolivar Coastal Fields has been described in detail recently.[20] The Gafsa Basin chert (Paleocene, Tunisia) was a gift from Professor Jean Trichet (Universite d'Orleans, France). The chert studied was a core section 261.3 m deep from the Redeyef SR15 core drilled in the Gafsa Basin about 60 km East of the town of Gafsa. The phosphatized Gafsa Metlaoui Basin was a marine gulf during the Paleocene, and the phosphatized series, which is 20m thick comprising nine phosphatized beds interstratified with non-phosphatized or poorly phosphatized sediments, includes the chert studied here. The bulk of the organic matter within the sample is of marine planktonic origin. Vitrinite reflectance studies gave values < 0.5%, indicative of an immature sediment. Plots of hydrogen index versus oxygen index confirmed the planktonic nature of the organic matter.[21]

2.2 Isolation Procedures

The extraction procedures for the isolation of high carbon number geoporphyrins from the Boscan oil, and the isolation of the metal free porphyrins from the Gafsa Basin are presented elsewhere.[22,23]

A novel series of vanadyl geoporphyrins were isolated from Boscan oil by chromatography on alumina and silica gel columns. The porphyrins were isolated from a grade II alumina column developing with hexane, hexane/toluene, toluene, toluene/CH_2Cl_2, CH_2Cl_2 and methanol. The porphyrin fraction eluted in 75% toluene/hexane. The concentrate was then chromatographed on silica gel eluting with hexane, CH_2Cl_2/hexane and CH_2Cl_2. The vanadyl porphyrins eluted in 20 to 50% CH_2Cl_2/hexane. The product was chromatographed on thin layer plates eluting with 12% CH_2Cl_2/toluene, and a red porphyrinic band was obtained with Rf 0.6-0.7.

2.3 Mass Spectrometry

Electron impact mass spectrometric (EIMS) and tandem mass spectrometric data were obtained on a triple stage quadrupole mass spectrometer (TSQ45 with INCOS data system, Finnegan MAT, San Jose California) with ion source conditions of 70eV electron energy, 0.3 mA emission current, and 190° C. MS/MS daughter spectra were obtained at a collision energy of 23.8 eV using nitrogen as the collision gas. [22]

3. RESULTS and DISCUSSION

3.1 MS/MS Analyses of Porphyrin Standards

The main feature of the EI MS/MS daughter spectra of the molecular ions of deuteroporphyrin-IX dimethyl ester, aetioporphyrin-III (1a) the tetra-n-pentyl porphyrin (1b), and the tetra-n-heptyl porphyrin (1c) was the predominance of cleavage β to the porphyrin macrocycle. [24] This trend is well illustrated in the daughter spectrum of the molecular ion of the tetra-n-pentyl porphyin (1b) . There are virtually no fragment ions besides the M-57 ion (Fig. 1). Thus the daughter spectra of the molecular ions of porphyrins furnish considerable information on the nature of the substitution patterns present. The metalloporphyrins behaved similarly. There were severe problems in obtaining reproducible CI MS/MS analyses of the porphyrins. In some cases there was substantial fragmentation into the individual pyrrolic components, as is observed in conventional CIMS analyses[8], but at other times little fragmentation occurred.

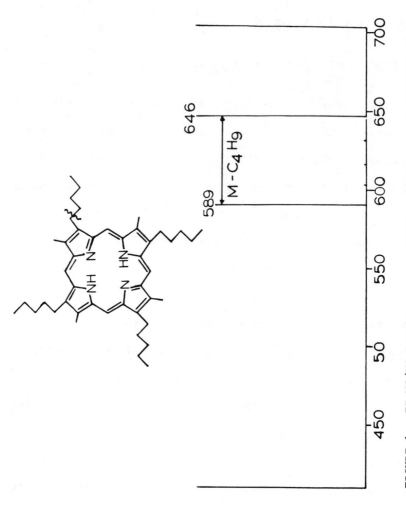

FIGURE 1   EI MS/MS Daughter ions of Molecular ion (m/z 646) of 1(b).

3.2 Analyses of High Carbon Number Geoporphyrins

The EIMS data for one of the predominantly aetio porphyrin
fractions are shown in Figure 2. The spectrum shows a wide
range of components (from at least C_{33}, m/z 492, to at least
C_{43}, m/z 632), but gives no data on the nature of the
individual components. The daughter spectra of the molecular
ions of the C_{36} aetio porphyrin m/z 534, and the C_{36} DPEP
component, m/z 532, which occurs in a more polar fraction,
illustrate the salient features of the analyses for all the
high carbon number species (Fig. 3). A more detailed
treatment of the data is presented elsewhere.[22,24]

The following inferences can be drawn from the spectra,
assuming that the geoporphyrin fragmentation patterns
resemble those of the porphyrin standards, and that the
porphyrins were derived from naturally-occurring
chlorophylls in a Treibs' type degradation pathway:

(a) There are a number of porphyrin isomers present in both
the DPEP and the aetio series.

(b) There is likely to be more than one site of extended
alkylation on the aetio porphyrin macrocycle. This is
the only way to account for the presence of porphyrins
with butyl and pentyl substituents as indicated by the

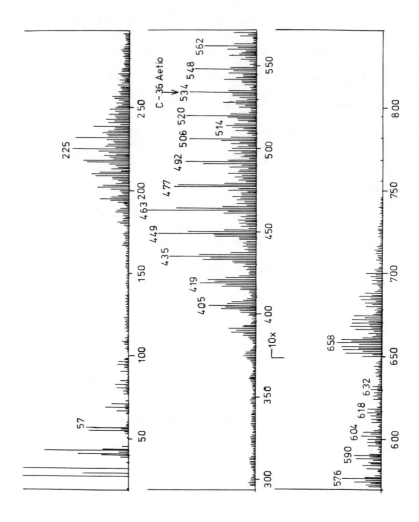

FIGURE 2.   EIMS of high carbon number aetio porphyrins of Boscan oil.

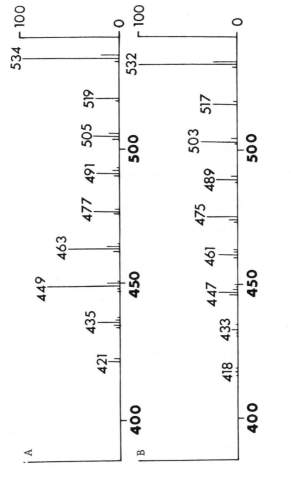

FIGURE 3.   Daughter ions of C_{36} Aetio (A) and C_{36} DPEP (B) porphyrin molecular ions from Boscan oil.

daughter ions m/z 491 and 489 in the spectra of the C_{36}
aetio and DPEP porphyrins respectively (Fig. 3)

(c) For all molecular ions studied, the DPEP species showed
less intense daughter ions than the corresponding aetio
components (Fig. 3). This difference in intensity is not
an inherent property of the DPEP porphyrins as it was
not observed when lower carbon number aetio and DPEP
porphyrins were compared.

(d) The differences in the distributions may indicate that
DPEP porphyrins with extended isocyclic rings, e.g. the
DPEP-6 and DPEP-7, species are more abundant than in
samples which are less mature.[5,8]

The high carbon number geoporphyrins in crude oil may
eventually prove very valuable as fingerprints for oil-oil
and oil-source rock correlation studies. Studies are being
initiated to evaluate this possibility.

3.3 Isolation of a Novel Class of Geoporphyrins

A suite of non-polar vanadyl porphyrins with mass
ranges corresponding to C_{36}-C_{42} benzo-DPEP porphyrins was
isolated from Boscan oil (see Experimental). The EIMS
spectrum of the mixture is shown in Figure 4. The daughter
spectrum of each molecular ion of the suite also showed no
fragment ions. The lack of fragment ions cannot be

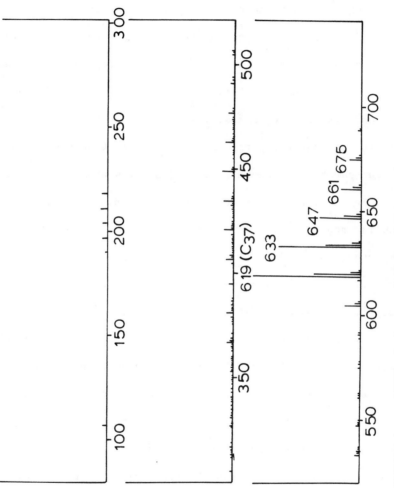

FIGURE 4.   EIMS of the novel benzo-DPEP porphyrin series of Boscan oil.

attributed to the properties of the benzo porphyrin macrocycle itself because analysis of benzo porphyrin standards bearing ethyl groups showed the expected β-cleavage (M-15 fragments) in both metal and metal-free forms.[25] Thus the geoporphyrins do not contain alkyl moieties which readily undergo β-cleavage i.e. they have no ethyl, or other alkyl substituents $>C_1$. The nature of these compounds is, as yet, a mystery. The major impediment to the solution is that the porphyrins are difficult to demetallate without substantial decomposition. Efforts are underway to develope an improved, milder demetallation procedure. Once this is achieved, it should be possible to identify the compounds, and to determine whether they are true benzo-DPEP porphyrins, or a new class of porphyrins which bear four isocyclic rings.

All known biologically-occurring chlorophylls have at least one ethyl substituent, which is absent in the entire suite of porphyins. Therefore one conclusion can be drawn – this porphyrin class cannot be derived from a Treibs' type degradation pathway in which only the chlorin functional groups undergo reaction, and the alkyl groups remain intact.

3.4 Analyses of Metal   Free Geoporphyrins   from Gafsa   Basin
    Chert

The distribution of metal free geoporphyrins from the   Gafsa
Basin differed   markedly   from   those   of   nickel   porphyrin
distributions from immature sediments.[1] The principle aetio
porphyrin components of the mixture were   aetioporphyrin-III
(1a), and   a   desethyl C_{30} porphyrin   which occur   in   many
immature   sediments.   However,   the   DPEP   distribution   was
unusual.   Along   with   the   expected
deoxophylloerythroetioporphyrin (2a),   an   isomeric C_{32} DPEP
was   also   abundant.   A C_{31} DPEP   with   an   unsubstituted β
-position was   isolated and   tentatively assigned   as   (2b).
This component is likely to   be a product of degradation   of
chlorophyll b (9). A C_{33} di-DPEP   (m/z 488) and C_{33} and C_{34}
benzo-DPEP (m/z 484, 498)   porphyrins were detected by   EIMS
in the C_{32} DPEP fraction.[23] These compounds are particularly
interesting because they have not been detected in   immature
sediments   before.   Thus   there   is   evidence that   benzo
porphyrins may   be   formed   quite early   in   the   diagenetic
pathway from the chlorophylls to the geoporphyrins.

4. CONCLUSIONS

Tandem mass spectrometry provides a very rapid, flexible and efficient method for the analyses of complex mixtures. It has proved particularly valuable in studying petroporphyrin concentrates which cannot be readily resolved into their individual components. MS/MS analyses of the high carbon number petroporphyrins reveal a complex mixture of isomers which may prove valuable as fingerprints in correlation studies. In addition, a novel series of porphyrins which cannot be formed by a Treibs' type degradation has been detected, and the first evidence for the formation of benzo porphyrins in the early stages of diagenesis has been obtained.

5. REFERENCES

1. E.W. BAKER and S.E. Palmer – Geochemistry of Porphyrins, The Porphyrins, Volume 1, edited by D.Dolphin, (John Wiley & Sons, New York, 1978) Chapter 11, pp.485-551.
2. S.E. PALMER and E.W. BAKER, Science, 201, 49-51 (1973).
3. E.W. BAKER, S.E. PALMER and W.Y. HUANG- Chlorin and Porphyrin Geochemistry of DSDP Leg 40 Sediments: Cape Verde Rise and Basin, Initial Reports of the Deep Sea Drilling Project, XL, edited by Y. Lancelot, E. Seibold et al.,(U.S. Government Printing Office, Washington, D.C., 1978) pp. 639-647
4. J.M.E. QUIRKE, G. EGLINTON and J.R. MAXWELL, J. Am. Chem. Soc., 101, 7693-7697 (1979).
5. G.A. WOLFF, M. MURRAY, J.R. MAXWELL, B.K. HUNTER and J.K.M. SANDERS, J. Chem. Soc., Chem. Commun., 1983, 922-924.

6. C.W.J. FOOKES, J. Chem. Soc., Chem. Commun., 1983, 1472-1474.

7. C.W.J. FOOKES, J. Chem. Soc., Chem. Commun., 1985, 706-708.

8. G.A. WOLFF, M.I. CHICARELLI, G.J. SHAW, R.P. EVERSHED, J.M.E. QUIRKE and J.R. MAXWELL, Tetrahedron, 40, 3777-3786 (1984).

9. J.M.E. QUIRKE and J.R. MAXWELL, Tetrahedron, 36, 3453-3458 (1980).

10. M.I. CHICARELLI, G.A. WOLFF and J.R. MAXWELL, J. Chem. Soc., Chem. Commun., 1985, 723-724.

11. A. TREIBS, Angew. Chemie, 49, 682-686 (1936).

12. E.W. BAKER, T.F. YEN, J.P. DICKIE, R.E. RHODES and L.F. CLARK, J. Am. Chem. Soc., 89, 3631-3639 (1967).

13. J.M.E. QUIRKE, G.J. SHAW, P.D. SOPER and J.R. MAXWELL, Tetrahedron, 36, 3261-3267 (1980).

14. B. DIDYK, Y.I.A. ALTURKI, C.T. PILLINGER and G. EGLINTON, Chem. Geol., 15, 193-208 (1975).

15. G.W. HODGSON, M. STROSHER and D.J. CASAGRANDE — Geochemistry of Porphyrins; Analytical Oxidation to Maleimides. Advances in Organic Geochemistry, 1971, edited by H.R. von Gaertner and and H. Wehner (Pergamon, Oxford, 1972), , pp.151-161.

16. B.M. DIDYK, B.R.T. SIMONEIT and G. EGLINTON, Org. Geochem., 5, 99-109 (1983).

17. F.W. MCLAFFERTY, Tandem Mass Spectrometry (John Wiley & Sons, New York, 1983).

18. H.D. HEDBERG, Bull. Amer. Petrol. Geol., 48, 1755-1803 (1964).

19. J.A. GRANSCH and E. EISMA — Geochemical Aspects of the Occurrence of Porphyrins in West Venezuelan Mineral Oils and Rocks. In Advances in Organic Geochemistry 1969, edited by G.D. Hobson and G.C. Speers, (Pergamon Press, Oxford, 1970) pp.69-86.

20. H. BOCKMEULEN, C. BARKER and P.A. DICKEY, Bull. Amer. Assoc. Petrol. Geol., 67, 242-270 (1983).

21. H. BELAYOUNI and J. TRICHET — Preliminary Data on the Origin and Diagenesis of the Organic Matter in the Phosphate Basin of Gafsa Tunisia, Advances in Organic Geochemistry 1981 edited by M. Bjorφy et al., (John Wiley & Sons, Chichester, 1983) pp.328-335.

22. J.V. JOHNSON, E.D. BRITTON, R.A. YOST, L.L. CUESTA and J.M.E. QUIRKE, Anal. Chem., 58, 1325-1329 (1986).

23. J.M.E. QUIRKE, T. DALE, J. TRICHET and H. BELAYOUNI, E.D. BRITTON and R.A. YOST, Org. Geochem., submitted.

24. J.M.E. QUIRKE, L.L. CUESTA, R.A. YOST, J.V. JOHNSON and E.D. BRITTON. Org. Geochem., in press.

25. J.M.E. QUIRKE, unpublished data.

INCORPORATION OF PETROPORPHYRINS INTO GEOCHEMICAL CORRELATION PROBLEMS

Padmanabhan Sundararaman, J. Michael Moldowan [x] and
Wolfgang K. Seifert [#] Chevron Oil Field Research Company, La Habra,
California 90631, and [x] P.O. Box 1627, Richmond, California
94802-0627. [#] Deceased 1985.

The first detailed comparison of hydrocarbon biomarker maturity parameters with two newly developed porphyrin parameters ($C_{32}DPEP/C_{30}ETIO$ and $C_{31}DPEP/C_{32}ETIO$) is described. Using crude oils from three different fields McKittrick field and San Joaquin, California, and Prudhoe Bay, Alaska, it is shown that porphyrin parameters are more sensitive to subtle maturity changes than are hydrocarbon biomarker parameters. Maturity differentiation by porphyrins is demonstrated for mature oils as well as heavily biodegraded oils. No one parameter, but a combination of porphyrin and hydrocarbon biomarker parameters, was able to provide conclusive results on oil maturity ranking in these basin studies.

1. INTRODUCTION

In recent years "biological markers" such as steranes, terpanes, mono- and triaromatized steroid hydrocarbons have been used to correlate oils and source rocks and to demonstrate relative maturities. Of the several commonly used hydrocarbon maturity parameters, only four, 22S/22R 17 α -homohopanes, 20S/20R 14 α ,17 α (H) steranes, 14 β ,17 β (H),20R/14 α ,17 α (H),20R steranes and mono-/triaromatized steroid hydrocarbons appear to be source independent. Even these parameters have certain disadvantages as described below.

The isomerization of the C-22 position of 17 α -homohopanes is extremely facile and equilibrium is reached at the beginning of or even before the petroleum generation window(Ro = 0.5%). For this reason, the 22S/22R 17 α -homohopane ratio is useful in demonstrating a low level of maturity for source rocks but rarely for oils.

The 20S/20R isomerization of 14 α ,17 α (H) steranes is slower than that of the C-22 position of homohopanes and extends into the petroleum generation window, thus making it a useful maturity parameter for oils. Even in this case equilibrium is reached fairly early. An exact correlation between these biomarker parameters and the oil generation window or even %Ro is not possible because the individual biomarker conversions are independently tied to heating rate and time of heating (MACKENZIE and McKENZIE 1983).

The epimerizations at the C-14 and C-17 position of steranes are slower than that at the C-20 position, and utilization of the 14 β ,17 β (H) to 14 α ,17 α (H) ratio as a maturation parameter extends to higher maturities, and can be used for more mature oils. The 14 β ,17 β (H) compounds may, in certain cases, migrate faster than the 14 α ,17 α (H) compounds (SEIFERT and MOLDOWAN, 1981) thereby superimposing a migration component. Recent work (ten Haven et. al., 1986) has also shown, in some cases the occurrence of early diagenetic 14 β ,17 β (H) steranes which complicates their use in maturity studies.

The mono-/triaromatized (MA/TA) steriod ratio extends still further into the oil window and has proven to be an excellent parameter for maturity evaluation. This parameter is even more likely to be affected by a migrational difference between its components (HOFFMANN et al., 1984).

Thus the source independent maturity parameters mentioned above have certain limitations and cannot be used universally for maturity assessments. Apart from these parameters, the rest of the hydrocarbon parameters have a superimposed source component (e.g., SEIFERT and MOLDOWAN, 1978) determined by the biological input and the conditions during early diagenesis.

The destruction or alteration of biomarkers in heavily biodegraded oils (SEIFERT et. al., 1984, and references cited therein) poses a limitation on their applicability in maturity studies. Recent work has even demonstrated biodestruction of mono- and triaromatized steroid hydrocarbons (WARDROPER et al., 1984). Porphyrins appear to be highly resistant to biodegradation (PALMER, 1983; BARWISE and PARK, 1983), thus making them potentially useful for determining the maturity of heavily biodegraded oils.

For this reason, it was decided to approach porphyrin analysis at the molecular level and ultimately develop geochemical correlations which reflect credible diagenetic precursor/product relationships.

1.1 Porphyrins as Maturity Parameters

Geological porphyrins are derived from chlorophyll and bacteriochlorophylls (OCAMPO et.al., 1986) via the scheme proposed by Treibs (1936). The porphyrins in petroleum occur as several different structural types, the major ones are DPEP and ETIO. The DPEP type porphyrins can be directly linked to chlorophyll "a", "b" and "c" and are the predominant components of low maturity sediments. With increasing thermal maturity, the DPEP components are thought to undergo cleavage of the isocyclic ring giving rise to the ETIO porphyrins (CORWIN, 1960). The proposed link between DPEP and ETIO porphyrins, however, has been challenged by Barwise and Roberts (1984). The more probable explanation according to these authors is that the ETIO porphyrins are generated during early diagenesis (Figure 1), and the reason for the decrease in the DPEP/ETIO ratio with maturity is that the DPEP components are destroyed faster than the ETIO components.

The ratio of DPEP/ETIO is still an indicator of thermal maturity of oils and sediments. Several workers (BAKER and LOUDA, 1983; MACKENZIE et al., 1980; BARWISE and PARK, 1983) have shown that the DPEP/ETIO ratio decreases with increasing maturity for oils and sediments.

The maturation of porphyrins probably follows two different pathways. Both DPEP and ETIO porphyrins can undergo benzylic clevage to produce lower carbon number DPEP and ETIO porphyrins. The second pathway involves hydrogenation of the metalloporphyrins to produce di-, tetra-, and higher hydroporphyrins. These hydrogenated metallo- porphyrins can undergo demetallation to produce free base porphyrins which decompose further. The work of Corwin and Wei (1962) supports this hypothesis. They found that the dissociation constants for magnesiumporphyrins are less than those for magnesium chlorins (dihydroporphyrins). Interestingly, closure of the isocyclic ring raises the dissociation constant (BUCHLER, 1975) making DPEP chlorins less stable than the ETIO chlorins, an observation which lends support to Barwise and Robert's hypothesis on the relative stability of DPEP and ETIO porphyrins.

Figure 1 suggests that C_{32} DPEP and C_{30} ETIO porphyrins are derived from the same precursor, i. e. chlorophyll "a". If the DPEP porphyrins decompose faster than the ETIO porphyrins, then the ratio C_{32} DPEP/ C_{30} ETIO should be a more accurate indicator of maturity than DPEP/ETIO. Recently Strom et al. (1984), published the structure of a C_{31} DPEP porphyrin Abelsonite. This, according to these authors probably arises from chlorophyll "a" by a microbially mediated vinyl scission (Fig. 2). Based on Figure 2, one would also expect C_{31} DPEP/ C_{29} ETIO to be a good indicator of maturity . We have determined the ratios of C_{32} DPEP/ C_{30} ETIO and C_{31} DPEP/ C_{29} ETIO from mass spectroscopy data and compared them with other hydrocarbon maturity parameters discussed above.

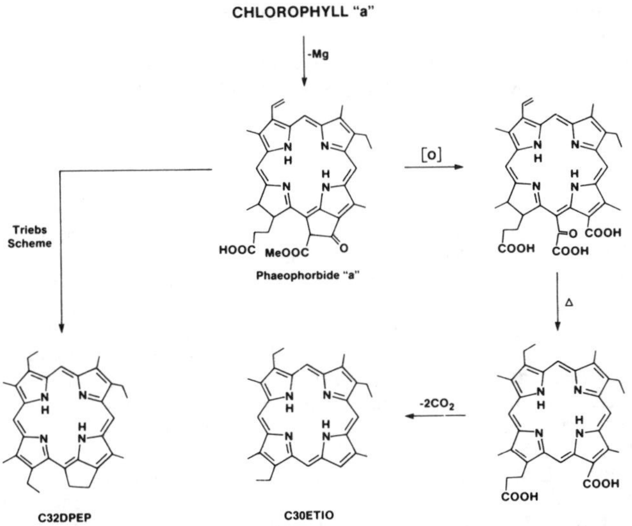

Figure 1   Proposed diagenetic pathway leading to the formation of $C_{30}ETIO$ and $C_{32}DPEP$ porphyrins from chlorophyll "a".

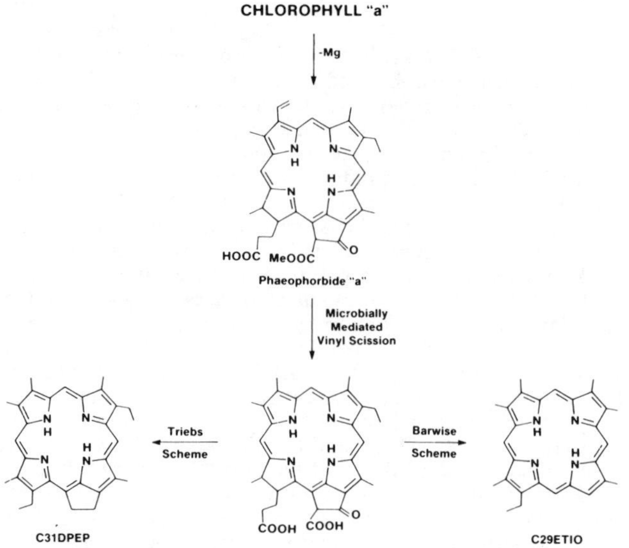

Figure 2 Proposed diagenetic pathway leading to the formation of $C_{29}ETIO$ and $C_{31}DPEP$ porphyrins from chlorophyll "a".

Several different isomers within each carbon number and type have been identified. Unfortunately probe mass spectrometry cannot distinquish these isomers thus the ratio C_{32} DPEP/ C_{30} ETIO and C_{31} DPEP/ C_{29} ETIO incorporates all of the DPEPs and ETIOs within a particular carbon number. Our recent work with the high performance liquid chromatography of vanadylporphyrins (SUNDARARAMAN and PETERS 1986) shows only some of the isomers within a single carbon number undergo changes with increasing maturation.

2: EXPERIMENTAL

Complete experimental details on isolation and purification have been published elsewhere (SUNDARARAMAN, 1986). The mass spectral data were obtained on a VG Micromass 7070H operating in the EI mode. Electron impact voltage was 70 eV. Sample introduction was by direct insertion probe; the temperature was programmed ballistically from abient to 350°C in 30 minutes. The spectrometer was scanned cyclically from m/z 10 to m/z 600 every 3 seconds. Data were acquired over the entire volatility range of the sample. The mass spectra were averaged and normalized using a Finnigan INCOS data system. The C_{32} DPEP/ C_{30} ETIO and C_{31} DPEP/ C_{29} ETIO ratios were determined from the intensities of m/z 541/515 and m/z 527/501 peaks, respectively.

Hydrocarbon data were obtained by gas chromatography- mass spectrometry either on a Finnigan or VG Micromass 7070H system. The GC column was DB-1 60m fused silica from J & W Scientific programmed at 2°C/minute from 150°C to 320°C. Hydrocarbon data were collected at least 3 times for each sample over a 2 year period. Absolute variations in the 20S/20R C_{29} -sterane ratios were + .03 and in the MA/TA-steroid ratios + 15%, but relative numerical order remained unchanged.

3. APPLICATIONS

3.1 McKittrick Field, California

The geology and geochemical relationship of the McKittrick oils (Carneros, Phacoides and Oceanic oils) has been described previously (SEIFERT and MOLDOWAN, 1978). The Carneros oils were shown to have source specific biomarker and carbon isotope ratios consistent with a common source. However inspection of the porphyrin parameters in Table I shows a large jump by a factor of 2 in the nickel/vanadyl porphyrin ratio for the 6480' oil. Based on our work (MOLDOWAN et al., 1986) on Toarcian shale from N. W. Germany we can conclude that this oil is probably derived from a source rock which was deposited in a relatively less reducing environment compared to the source for the other oils. Rearranged/regular sterane and pristane/phytane ratios were also shown in the Toarcian sediments to vary sympathetically with Ni/(Ni + V) porphyrin ratios depending on the environment of deposition. Thus, an elevated pristane/phytane ratio (Table I) plus an elevated rearranged/regular sterane ratio in the 6480 oil (SEIFERT AND MOLDOWAN, 1978, Table 9) are in agreement with a less reducing depositional environment for the source of that oil.

TABLE I   Maturation parameters of Carneros oils from McKiitrick Field, California

Depth, Ft	Porphyrins				Hydrocarbons			Bulk
	Nickel/ Vanadyl	DPEP/ ETIO	$C_{32}D/C_{30}E^2$	$C_{31}D/C_{29}E^2$	Prist/Phyt	MA/TA^1	$20S/20R^3$	API
5655	4.3	1.18	1.40	1.21	1.45	0.18	0.53	32.8
5970	4.7	1.37	1.63	1.35	1.45	0.20	0.55	35.5
7194	4.3	0.63	0.84	0.70	1.44	0.21	0.58	33.2
5872	4.5	0.66	0.95	1.10	1.46	0.23	0.54	35.4
6480	8.9	1.18	1.38	1.14	1.56	0.26	0.58	41.1

1 Mono/Triaromatized Steroids, $C_{27} - C_{29}/C_{26} - C_{28}$.
2 D = DPEP; E = ETIO.
3 $5\alpha,14\alpha,17\alpha$ -steranes, C_{29}.

Substantial differences are found in the DPEP/ETIO ratio of the demetallated nickelporphyrins in the Carneros oils (Table I). The data indicate that the McKittrick oils from 7194' and 5872' depths are of equal and highest maturity among the Carneros group. However, a subtle maturity distinction between these two oils is revealed by C_{32} DPEP/ C_{30} ETIO and C_{31} DPEP/ C_{29} ETIO parameters. In contrast the hydrocarbon maturity data for the Carneros oils (Table I, MA/TA and 20S/20R) barely range outside the limits of precision of the measurements (see experimental section). This points out a greater sensitivity of porphyrins to slight maturity differences among oils.

For another example, the Phacoides and Oceanic oils in Table II are from geologically older sources in the same basin and have been shown to be more mature than the Carneros oils (SEIFERT and MOLDOWAN, 1978). The greater maturity of this source group is indicated here by lower MA/TA steroid ratios and higher 20S/20R C29-sterane ratios which occupy a narrow distribution among the five oils. The narrowly ranging API gravities are lower (falsely indicating lower maturity) than those of the Carneros oils, which illustrates the problem of comparing source dependent maturity parameters between oils from different sources, even in the same basin.

TABLE II   Maturation parameters of oils from McKittrick Field, California

Formation		Porphyrins		Hydrocarbons		Bulk
Age	Depth	$C_{32}D/C_{30}E^2$	$C_{31}D/C_{29}E^2$	MA/TA^1	$20S/20R^3$	API
Carneros/L.Mio.		0.84-1.63	0.70-1.35	0.20-0.26	0.53-0.58	32-41
Phacoides/Olig.	7828	0.71	0.57	0.15	0.59	34.4
Phacoides/Olig.	8210	0.93	0.80	0.15	0.63	32.6
Phacoides/Olig.	9058	0.59	0.51	0.14	0.65	32.5
Oceanic/Olig.	8858	0.54	0.58	0.16	0.58	31.3
Oceanic/Olig.	8834	0.91	0.91	0.17	0.62	31.7

1 Mono/Triaromatized Steroids, $C_{27} - C_{29}/C_{26} - C_{28}$.
2 D = DPEP; E = ETIO.
3 $5\alpha,14\alpha,17\alpha$ -steranes, C_{29}.

Unlike the hydrocarbon data, we observe significant differences in the C_{32} DPEP/ C_{30} ETIO and C_{31} DPEP/ C_{29} ETIO porphyrin ratios for the Phacoides and the Oceanic group of oils. These differences allow us to arrange these oils in order of increasing maturity. Thus based on porphyrins the sequence of maturity is 8210' < 7828' < 9058' for the Phacoides oils and 8834' < 8858' for the Oceanic oils. Although the Phacoides/Oceanic group is also shown to be more mature than the Carneros group by porphyrin ratios, there is one discrepancy. The most mature 7194' Carneros oil overlaps into the geologically older sourced oil group, but this is not supported by hydrocarbon "source independent" maturity parameters. This may be an indication of some residual source dependency in the DPEP/ETIO ratios, a relic of early diagenetic ETIO formation (i.e. Figs. 1 and 2). For higher maturity oils this effect may be obliterated due to higher maturity response of the DPEP/ETIO ratios.

3.2 Prudhoe Bay, Alaska

For background information on the geology and source of these oils, the reader is referred to previous publications (SEIFERT et al., 1980; 1983). This earlier work indicates that the three oils described in Table III are derived at least in part from a deep post-Neocomian source, but the Sadlerochit and Lisburne oils have co-sources. Hence, although there is no one-to-one source relationship among the oils, there is enough similarity that a direct comparison of maturity parameters is possible.

Inspection of the hydrocarbon maturity parameters in Table III clearly shows that Kuparuk River oil is the least mature of the three oils. The 14 β ,17 β (H),20R/14 α 17 α (H),20R sterane ratio shows that the conversion of the biological precursor (14 α ,17 α (H),20R to the geological product (14 β ,17 β (H),20R) has proceeded to a lesser extent in the Kuparuk River oil compared to Sadlerochit and Lisburne oils. On the other hand, in the Sadlerochit and Lisburne oils the ratio 14 β ,17 β (H),20R/14 α ,17 α (H),20R C_{29} -steranes does not

distinquish the maturity differences between them. This conversion has reached
a value which appears to be close to a maximum value observed in many mature
oils. There may be no absolute equilibrium value for oils as the 14 β ,17 β
(H),20R/14 α ,17 α (H),20R ratio appears to be highly variable with temperature
(Van Grass et al., 1982). The MA/TA ratios are too close to distinquish maturity
differences between these oils.

TABLE III   Maturation parameters of oils from Prudhoe Bay, Alaska

Reservoir		Porphyrins		Hydrocarbons		Bulk
Formation/Age	Depth(Ft)	$C_{32}D/C_{30}E^2$	$C_{31}D/C_{29}E^2$	MA/TA [1]	$14\beta20R/5\alpha20R$ [3]	API
Sadlerochit/						
Permo-Triassic	9000	0.19	0.16	0.07	1.8	25.3
Lisburne/						
Mississippian	10000	0.37	0.40	0.10	2.0	24.3
Kuparuk River/						
L.Cretaceous	7000	0.74	0.87	0.12	1.4	26.6

[1] Mono/Triaromatized Steroids, $C_{27} - C_{29}/C_{26} - C_{28}$.
[2] D = DPEP; E = ETIO.
[3] C_{29} -Steranes

Based on the two porphyrin maturity parameters (i.e. C_{32} DPEP/ C_{30} ETIO
and C_{31} DPEP/ C_{29} ETIO) the order of maturity among the three oils is
Sadlerochit > Lisburne > Kuparuk River. The differences in the porphyrin ratios are much larger than the change in MA/TA-steroid ratio, indicating again a
greater sensitivity of porphyrins to maturation. The result also shows that the
proposed porphyrin maturity parameters are effective in evaluating more mature
oils than can be evaluated using sterane ratios.

Biodegraded Oils, California

The effects of biodegradation on various biological markers have been discussed
in numerous recent publications (SEIFERT et al., 1984; WARDROPER et al.,
1984; and references cited therein). Heavy biodegradation results in alteration
or destruction of steranes and terpanes, thereby limiting their potential use in
maturity correlations for biodegraded oils.
      Sometimes, however, the disappearance of the series of regular hopane
epimers during biodegradation is accompanied by the simultaneous appearance
of a series of desmethylated hopanes (REED, 1977; SEIFERT and
MOLDOWAN, 1979).
      In the case of the San Joaquin Basin oils it was shown that
25,28,30-trisnorhopane is formed through diagenesis and in heavily biodegraded
oils by bacterial alteration of 28,30- bisnorhopane (MOLDOWAN et al., 1984).
In that report these 25-demethylated hopane analogues (RULLKOTTER and
WENDISCH, 1982) were used in determining the order of maturity of two groups
of San Joaquin, California, oils. Specifically, the percentage of 17 α (H), 18 α (H),
21 β (H)-25,28,30-trisnorhopane in the sum of itself plus its 17 β (H), 21 α (H)
and 17 β (H), 21 β (H) epimers was found to decrease with increasing maturity.
The six oils are geochemically similar and may have a stratigraphically equivalent
source as discussed previously (MOLDOWAN et. al., 1984). These data along
with the other hydrocarbon parameters and porphyrin maturity parameters, are
listed in Table IV.

The shallow oils 214 and 60 were shown previously (SEIFERT and MOLDOWAN, 1979) to have undergone complete destruction of paraffins, isoprenoids and regular steranes. The trisnorhopane parameter groups these oils in the following maturity order:

$$61 < 214, 60 < 36, 274$$

This parameter is unable to clearly distinquish the maturity differences between 214 and 60, and between 36 and 274. The aromatized steroids (MA/TA ratio, Table IV), which survive biodegradation in this case, distinquish only 61 as the least mature oil and the differences among all the oils are small.

TABLE IV   Maturation parameters of biodegraded oils from the San Joaquin Valley, California

Sample	Depth	Porphyrins		Hydrocarbons		
		$C_{32}D/C_{30}E$ [3]	$C_{31}D/C_{29}E$ [3]	MA/TA [1]	$C_{27}TNH$ [2]	$20S/20R$ [4]
61	4378	3.18	3.06	0.23	75	0.33
214	1000	2.17	2.36	0.19	67	-
36	4900	1.45	1.71	0.19	52	0.40
60	1585	1.35	1.63	0.17	64	-
274	3032	1.36	1.50	0.16	50	0.40

[1] Mono/Triaromatized Steroids, $C_{27} - C_{29}/C_{26} - C_{28}$.
[2] 25,28,30 - Trisnorhopane C_{27}, Percentage of D/E Cis.
[3] D = DPEP; E = ETIO.
[4] $5\alpha, 14\alpha, 17\alpha$ -steranes, C_{29}.

Vanadylporphyrins, being highly aromatic, survive biodegradation (PALMER, 1983; BARWISE and PARK, 1983) which makes them extremely useful biological marker compounds in geochemical correlations. The two porphyrin maturity parameters (C_{32} DPEP/ C_{30} ETIO and C_{31} DPEP/ C_{29} ETIO, Table IV) clearly show the 61 oil to be the least mature of the five oils; and they can be used to distinquish the maturity differences between 214 and 60 oils, the former one being less mature. Interestingly, the three oils 36, 60 and 274 appear to be of equal maturity by the porphyrin parameters.

Thus, the overall sequence of maturity predicted by the two porphyrin parameters agrees with the conclusion derived from the hydrocarbon parameters. In addition, the porphyrins bring to light maturity relationships which would not be evident based on the hydrocarbon parameters alone.

4. CONCLUSIONS

The two new porphyrin parameters (C_{32} DPEP/ C_{30} ETIO and C_{31} DPEP/ C_{29} ETIO) have been developed, and their potential use as indicators of thermal maturity has.been examined. This is the first detailed comparison of hydrocarbon biomarker maturity parameters with porphyrin maturity parameters. As a vehicle for this comparison, maturation parameters of crude oils from three locations, Prudhoe Bay, Alaska, McKittrick and San Joaquin, California, are described. Porphyrin parameters are shown to be more sensitive to subtle maturity changes than are the hydrocarbon parameters. Their use in estimating the maturity of moderately mature, as well as heavily biodegraded oils is demonstrated. It is apparent that no one parameter can lead to conclusive results, but a multiparameter approach using complementary parameters is the best way of solving geochemical problems.

4. ACKNOWLEDGEMENTS

The authors thank E. J. Gallegos for his assistance with MS instrumentation, Ms M. M. Pena for the porphyrin experimental work, and Ms. C. Y. Lee for hydrocarbon data handling.

REFERENCES

BAKER E. W. and LOUDA J. W. (1983) Thermal aspects in chlorophyll geochemistry. In Advances in Organic Geochemistry 1981 (eds. M. Bjoroy et. al.), pp 401-421. Wiley, Chichester.

BARWISE A. J. G. and PARK P. J. D. (1983) Petroporphyrin fingerprinting as a geochemical marker. In Advances in Organic Geochemistry 1981 (eds. M. Bjoroy et. al.) pp 668-674. Wiley, Chichester.

BARWISE A. J. G. and ROBERTS I. (1984) Diagenetic and catagenetic pathways for porphyrins in sediments. In Advances in Organic Geochemistry (eds. P. A. Schenck, J. W. de Leeuw and G. W. M. Lijmbach) Organic Geochemistry 6, pp 167-176, Pergamon Press, Oxford.

BUCHLER J. W. (1975) Static coordination chemsitry of metalloporphyrins. In Porphyrins and Metalloporphyrins (ed. K. M. Smith), pp 205-206. Elsevier, Amstradam.

CORWIN A. H. (1960) Petroporphyrins. Proc. 5th World Petroleum Congress, New York, Paper V-10, pp 119-129.

CORWIN A. H. and WEI P. E. (1962) Stability of magnesium chelates of porphyrins and chlorins. J. Org. Chem. 27, 4285-4290.

HOFFMANN C. F., MACKENZIE A. S., LEWIS C. A., MAXWELL J. R., VANDENBROUCKE M., DURAND B. and OUDIN J. L. (1984) A biological marker study of coals, shales and oils from the Mahakam Delta, Kalimantan, Indonesia. Chem. Geol. 42, 1-23.

LOUDA J. W. and BAKER E. W. (1981) Geochemiatry of tetrapyrrole, carotenoid and perylene pigments in sediments from the San Miguel Gap (Site 467) and Baja California borderland (site 471), Deep Sea Drilling Project Leg 63. In Initial Reports of the Deep Sea Drilling Project-LXIII, pp 785-818. U. S. Govt. Printing Office, Washington.

MACKENZIE A. S. and McKENZIE D. (1983) Isomerization and aromatization of hydrocarbons in sedimentary basins formed by extension. Geol. Mag., 120, pp 417-528.

MACKENZIE A. S., QUIRKE J. M. E. and MAXWELL J. R. (1980) Molecular parameters of maturation in the Toarcian Shales, Paris Basin, France-II. Evolution of metalloporphyrins. In Advances in Organic Geochemistry 1979 (eds. J. R. Maxwell and A. G. Douglas) pp 239-248. Pergamon Press, Oxford.

MACKENZIE A. S., LAMB N. A. and MAXWELL J. R. (1982) Steroid hydrocarbons and the thermal history of sediments. Nature 295, 223-226.

MOLDOWAN J. M., SEIFERT W. K., ARNOLD E. and CLARDY J. (1984) Structure proof and significance of stereoisomeric 28, 30-bisnorhopanes in petroleum and petroleum source rocks. Geochim. Cosmochim. Acta 48, 1651-1661.

MOLDOWAN J. M., SUNDARARAMAN P., and SCHOELL M., (1986) Sensitivity of biomarker properties to depositional environment and/or source input in the Lower Toarcian of SW-Germany. In Advances in Organic Geochemistry 1985 Organic Geochemistry 10, 915-926.

OCAMPO R., CALLOT H. J., AND ALBRECHT P. (1985) Occurrence of Bacteriopetroporphyrins in oil shale. J. Chem. Soc., Chem. Commun., 200-201.

PALMER S. E. (1983) Porphyrin distribution in degraded and nondegraded oils from Columbia. Abstracts, 186th American Chemical Society National Meeting, Washington, D. C., U. S. A., August 1983.

REED W. E. (1977) Molecular composition of weathered petroleum and comparison with its possible source. Geochim. Cosmochim. Acta 41, 237-247.

RULLKOTTER J. and WENDISCH D. (1982) Microbial alteration of 17 α (H) hopanes in Madagascar asphalts: removal of C-10 methyl group and ring opening. Geochim. Cosmochim. Acta 46, 1545-1553.

SEIFERT W. K. and MOLDOWAN J. M. (1978) Applications of steranes, terpanes and monoaromatics to the maturation, migration and source of crude oils. Geochim. Cosmochim. Acta 42, 77-95.

SEIFERT W.K. and MOLDOWAN J. M. (1979) The effect of biodegradation on steranes and terpanes in crude oils. Geochim. Cosmochim. Acta 43, 111-126.

SEIFERT W. K. and MOLDOWAN J. M. (1981) Palereconstruction by biological markers. Geochim. Cosmochim. Acta 45, 783-794.

SEIFERT W. K., MOLDOWAN J. M. and JONES R. W. (1980) Application of biological marker chemistry to petroleum exploration. Proc. 10th World Petroleum Congress, Bucharest, Romania, September 1979, paper SP8, pp 425-440, Heyden.

SEIFERT W. K., CARLSON R. M. and MOLDOWAN J. M. (1983) Geomimetic synthesis, structure assignment and geochemical correlation application of monoaromatized petroleum steroids. In Advances in Organic Geochemistry 1981 (eds. M. Bjoroy et al.) pp 710-724, Wiley, Chichester.

SEIFERT W. K., MOLDOWAN J. M. and DEMAISON G. J. (1984) Source correlation of biodegraded oils. In Advances in Organic Geochemistry 1983 (eds. P. A. Schenck, J. W. de Leeuw and G. W. M. Lijmbach) Organic Geochemistry 6, 633-643, Pergamon Press, Oxford.

STORM C. B., KRANE J., SKJETNE T., TELNAES N., BRANTHAVER J. F. and BAKER E. W. (1984). The structure of Abelsonite. Science 223, 1075-76.

SUNDARARAMAN P. (1985) High-performance liquid chromatography of vanadyl porphyrins. Analytical Chemistry 57, 2204-2206.

ten HAVEN H. L., de LEEUW J. W., PEAKMAN T. M. and MAXWELL J. R. (1986) Anomalies in steroid and hopanoid maturity indices. Geochim. Cosmochim. Acta 50, 853-855.

TRIEBS A. (1936) Chlorophyll and Hemin derivatives in organic materials. Angew. Chem. 49, 682-686.

VAN GRASS G, BAAS J. M. A., VAN de GRAFF B, and de LEEUW J. W. (1982) Theoretical organic geochemistry. I. The thermodynamic stability of several cholestane isomers calculated by molecular mechanics. Geochim Cosmochim Acta 46, 2399-2402.

WARDROPER A. M. K., HOFFMAN C. F., MAXWELL J. R., BARWISE A. J. G., GOODWIN N. S. and PARK P. J. D. (1984). Crude oil biodegradation under simulated and natural conditions-II. Aromatic steroid hydrocarbons. In Advances in Organic Geochemistry 1983 (eds. P. A. Schenck, J. W. de Leeuw and G. W. M. Lijmbach) Organic Geochemistry 6, 605-617, Pergamon Press, Oxford.

A BIOLOGICAL MARKER STUDY OF A TERTIARY SPHAGNUM COAL FROM YUNAN PROVINCE, THE PEOPLE'S REPUBLIC OF CHINA

JIAMO FU [1], GUOYING SHENG [1], SIMON C. BRASSELL [2,3], GEOFREY EGLINTON [2], ANN P. GOWER [2], DEYU CHEN [1] and DEHAN LIU [1]

        This paper is a preliminary study of newly-discovered sphagnum brown coal from the Jingsuo basin, Yunan province, China. This coal is quite rich in montan wax reflecting the predominance of higher n-alkane homologues and aromatic triterpenoid components. The aromatic hydrocarbons are mainly composed of pentacyclic and tetracyclic di-, tri and tetraaromatic compounds, which are obviously diagenetically related to naturally-occurring higher plant-derived triterpenoids in the biosphere. Given the low maturity of the brown coal, it is evident that progressive aromatization of higher plant-derived triterpenoids can start at a very early stage of diagenesis.

----

[1] Institute of Geochemistry, Academia Sinica, Guiyang, Guizhou Province, The People's Republic of China.

[2] Organic Geochemistry Unit, School of Chemistry, University of Bristol, Cantock's Close, Bristol BS8 1TS, United Kingdom.

[3] Present Address: Department of Geology, School of Earth Sciences, Stanford University, Stanford, CA 94305, U.S.A.

INTRODUCTION

In the field of molecular organic geochemistry the
appreciation and utilization of information contained within
biological marker distributions for the characterization,
differentiation and assessment of depositional environments
is sparsely documented (6,7), although significant advances
have been made in recent years (21, 24). An improved understanding of the potential of biological markers in paleoenvironmental assessment demands that a number of long-term
requirements be addressed and met. For example, it is necessary to establish and accumulate knowledge of the presence
of characteristic biological markers, or biomarker skeletons
in specific organisms (e.g., 4), and to relate such information to the occurrence of diagnostic biomarkers in sediments.
Such objectives are aided by studies of sediments with limited
or highly specific contributions of organic matter.

Sphagnum peats occur in many locations both within and
outside China, but sphagnum coal was discovered only
recently in Yunan province, China. It is believed to have
been deposited in an upland swamp environment dominated by
sphagnum input (Lujie and Zhang Xiuyi, personal communication). In this paper, a preliminary study of three
sphagnum brown coal samples was undertaken to explore and
comprehend the significance of biomarker compounds that may
be diagnostic of terrigenous higher plant inputs, especially
of sphagnum. In addition, comparison of the biomarker
characteristics of this sphagnum coal with their distributions in other brown coals and lignites (e.g., 9, 14, 24, 6, 36)
aids recognition of any distinguishing features specific
to sphagnum coal.

The sphagnum brown coals occur in two small intermoun-
tainous basins. These basins, Jingsuo and Qifu, are located

in the vicinity of Xundian and Qujing counties, respectively,
some 100 km northeast of Kunming, capital of Yunan province
(Fig. 1). The samples studied herein were collected from
Jingsuo basin which contains Middle to Upper Pliocene
inland lacustrine-swamp sediments, up to 60-130 m thick.
These deposits lie unconformable on Lower Palaeozoic clastic
and carbonate rocks, and contain one major coal layer up to
26 m thick (in outcrop). In the upper part of this coal
layer there are 19 sublayers of sphagnum coal, varying in
thickness from 8 to 93 cm. Geological and geochemical data
suggest that the sphagnum coal was deposited in a nutrition-
poor upland swamp environment. Its proportion of CaO + MgO
to SiO_2 + Al_2O_3 is low, only 2 to 3%.

EXPERIMENTAL

The samples were all extracted by Soxhlet (Ch_2Cl_2/MeOH
or toluene/MeOH). Their aliphatic and aromatic fractions
were isolated by standard TLC procedures and analysed by
gas chromatography (GC) and computerized gas chromatography-
mass spectrometry (C-GC-MS), according to previously published methods (21, 22, 27, 12).

During the development of sample JS-O on a TLC plate,
a sky-blue color band was observed under UV fluorescence.
This band was collected by solvent extraction (CH_2/Cl_2).
UV spectrometry indicated that it contained tri- and tetraaromatic compounds, as confirmed, subsequently, by GC and
C-GC-MS. Biological marker compounds were identified by
reference to standard compounds, literature spectra or by
spectral interpretation.

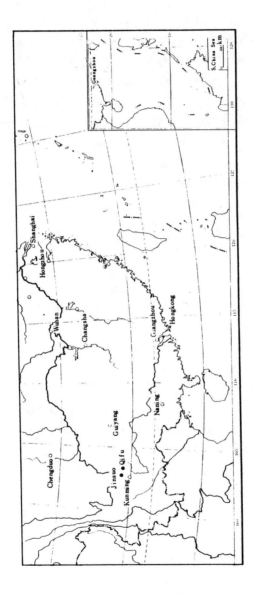

FIG. 1: Geographical map of South-West China showing the locations of the sphagnum brown coal deposits in the Jinsuo and Qifu Basins within Yunan Province.

RESULTS

ORGANIC PETROGRAPHY, IR AND ESR MEASUREMENT: ORGANIC
ELEMENTAL COMPOSITION AND SOM CONTENT

The sphagnum coal is mainly composed of sphagnum-
textinite (72-85%) with a low amount of sphagnum-fusinite
(Lujie, personal communication). Its wax content is high,
up to 13-15% (water free). The reflectance of the humulite
from the sphagnum brown coal has a R_{ran} value of 0.24%
indicating that the sphagnum coal lies within the soft lignite stage of maturity. This result compares well with ESR
measurement of the sphagnum coal (ESR concentrations of
0.13×10^{19}/g and 0.12×10^{19}/g for samples JS-1 and JS-2,
respectively).

Determination of the organic elemental content of the
samples (Table 1) provided H/C and O/C ratios for the sphagnum coal which are typical of type II kerogen. The soluble
organic matter (SOM) content of the samples is high, with
values of 16.32% and 31% for SOM/coal and SOM/C_{org}, respectively. The SOM is composed of 10% aliphatic hydrocarbons,
17% aromatic hydrocarbons and 73% non-hydrocarbons plus
asphaltenes. IR analysis shows dominant peaks for the SOM
at 2918 cm^{-1} and 2850 cm^{-1} which correspond to CH_2 and CH_3
stretching vibrations.

The high content of sphagnum-textinite in the sphagnum
coal and its high SOM values demonstrate that it possesses
good petroleum source potential.

ALIPHATIC HYDROCARBON BIOMARKERS

N-alkanes are the predominant components of the aliphatic hydrocarbon fraction (Fig. 2a), exhibiting a marked
odd/even predominance in the C_{21} to C_{35} region (OEP values
of 3.63 for JS-0 and 5.1 for JS-1) and major peaks at C_{31}

TABLE 1   ORGANIC ELEMENTAL COMPOSITION OF SPHAGNUM BROWN COAL SAMPLES

Sample Code	C % Weight	C % Atomic	H % Weight	H % Atomic	O % Weight	O % Atomic	N % Weight	N % Atomic	S % Weight	S % Atomic	H/C Atomic	O/C Atomic	N/C Atomic
JS-0	52.35	4.36	5.66	5.66	–	–	–	–	0.75	0.023	1.30	–	–
JS-1	52.39	4.37	6.51	6.51	23.97	1.50	0.78	0.055	–	–	1.49	0.34	0.013
JS-2	56.06	4.67	6.19	6.19	32.14	2.10	0.79	0.056	–	–	1.33	0.43	0.012

* Sample JS-0 was analysed in Bristol; samples JS-1 and JS-2 were investigated in Guiyang.

– not analysed.

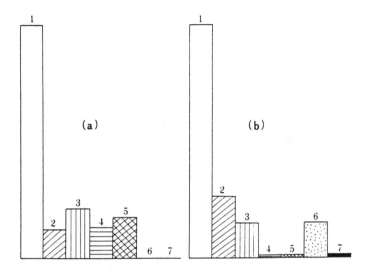

Fig. 2: Relative concentrations of biological markers in
        sample JS-O. A. Aliphatic hydrocarbon fraction:
        1. n-alkanes; 2. acylic isoprenoids; 3. hopanoids;
        4. triterpenes; 5. diterpenoids; 6. sesquiterpenoids;
        7. steranes; B. aromatic hydrocarbon fraction:
        1. tetraaromatic pentacyclics; 2. triaromatic
        pentacyclics; 3. diaromatic tetracyclics; 4. aroma-
        tic diterpenoids; 5. aromatic steroids; 6. monoarom-
        atic triterpenoids; 7. benzene, naphthalene, phenan-
        threne, and their alkyl derivatives.

or C_{29} (Fig. 3). Lower molecular weight n-alkanes ($<C_{21}$)
are minor constituents, hence the n-alkane distribution
pattern is not significantly bimodal. C_{16} and C_{18}-C_{20}
regular isoprenoid hydrocarbons were recognized; pristane
is the major compound of sample JS-0 which has a Pr/Ph ratio
of 3.33, significantly higher than the values ($<$unity) for
the other samples. No higher molecular weight acyclic isoprenoid alkanes were detected.

In addition to n-alkanes the sphagnum coal possesses
major amounts of triterpenoid and diterpenoid hydrocarbons
(Fig. 2a). The diterpenoid hydrocarbons detected include
a number of C_{19} and C_{20} diterpenes and diterpanes; abietane
(I), fichtelite (II) and isopimarane (III) were tentatively
identified by comparison of their spectra with literature
data (20). The distribution of hopanes is illustrated by the
m/z 191 mass chromatogram for sample JS-0 (Fig. 4). Hopanes
comprise C_{27} - C_{32} components, including 17α(H), 21β(H),
17β(H), 21α(H) and 17β(H), 21β(H) isomers with the last
dominant. No hopanes were detected, but peaks 1 and 2 of
Figure 4 were assigned as tarax-14-ene (IV) and olean-12-ene
(V), respectively, on the basis of their mass spectra and
retention behavior (18, 1). No steranes were detected in
the samples.

AROMATIC HYDROCARBON BIOMARKERS

Aromatic hydrocarbon fractions of all three samples
are dominated by tri- and tetraaromatic pentacyclic compounds (Fig. 2b). The tri - and tetraaromatic compound mixture of sample JS-0 was isolated by TLC and confirmed by its
visible-UV spectra, having absorbance bands at 245 nm and
250 nm. Five major components in this mixture (Fig. 5)
were identified by GC-MS; their structural assignments are

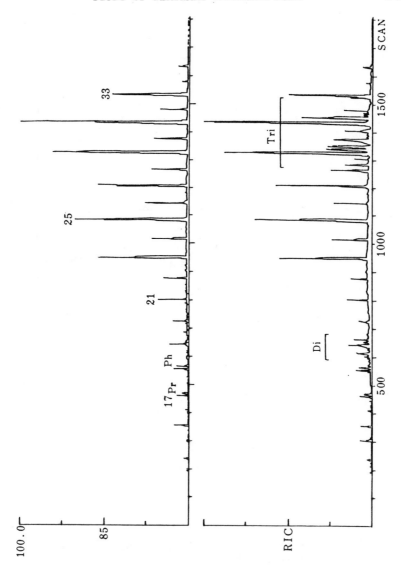

Fig. 3: Total ion reconstruction trace (a) and m/z 85
chromatogram (b) of aliphatic fraction of sample
JS-1. 17,25,31,$\underline{n}$-C$_{17}$, C$_{25}$, and C$_{31}$ alkanes; Pr,
pristane; Ph, phytane; Tri, pentacyclic triterpen-
oids; Di, tricyclic diterpenoids.

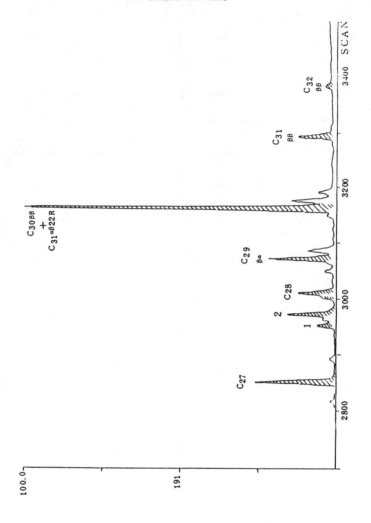

Fig. 4: m/z 191 chromatogram illustrating the distribution
of hopanes and triterpenes among the aliphatic
hydrocarbons in sample JS-O. Major hopane components are designated with their carbon numbers and
C-17, C-21 stereochemistry. Peaks 1 and 2 are
taraxer-14-ene and olean-12-ene, respectively; note
that m/z 191 is a minor ion in their spectra, hence
their abundance relative to the hopanes is significantly greater than it appears in this chromatogram.

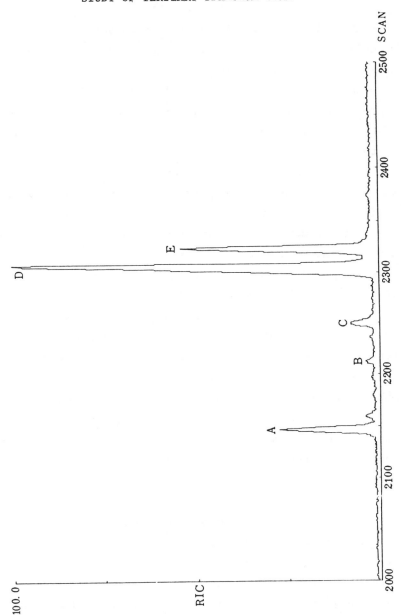

Fig. 5: Total ion reconstruction trace of tri- and tetraaromatic hydrocarbon fraction from sample JS-0.
Structural assignments of components A-E are given
in Table 2.

given in Table 2.(9). All these compounds have been previously reported from brown coal and marine, deltaic and
lake sediments (14, 31, 19, 34, 3, 33, 13). Three of these
five compounds (VI-X) were the only major components of the
aromatic hydrocarbons on sample JS-1 (Fig. 6). The presence
of two peaks for A probably reflects its occurrence as a
mixture of 14bα(H)- and 14bβ(H) isomers (9).

Other classes of aromatic components are less abundant,
but still important. One class has been reported previously.(34) in recent sediments and includes three tetracyclic
diaromatic compounds (XI-XIII; Fig. 7 and Table 2). Another
class contains isomers of an unknown compound, having a
base peak at m/z 158 (145) and a molecular ion at m/z 376
(Fig. 6 and Table 2; peaks K and L).

In addition, minor components were detected, including
dehydroabietant (XIV, 29), and alkylated derivatives of
benzene, naphthalene and phenanthrene. Two C_{29} monoaromatic steroidal compounds were identified, both of which
exhibited base peaks of m/z 211 and molecular ions of m/z
394 (Fig. 7 peaks I and J). From previous reports (17, 16),
their structures were tentatively assigned as A-ring steroid
and B-ring anthrasteroid monoaromatics (Table 2).

DISCUSSION

Sphagnum is an unusual type of brown coal which is rich in
wax and montan wax. The coal possesses significant amounts
of soluble organic matter, high ratios of aliphatic HC/
total extract and n-alkane/aliphatic HC, as well as a significant proportion of higher plant derived n-alkanes with
marked odd-over-even predominance. These characteristics
may reflect the high content of montan wax in sphagnum
coal.

TABLE 2   POLYCYCLIC AROMATIC HYDROCARBONS IN SPHAGNUM BROWN COAL

Component Peak	Proposed Assignment	Mass Spectra	Literature Reference Data	Proposed Structure (See Appendix)
A	2,2,4a,9-tetramethyl-1,2,3,4,4a,5,6,14b-octahydro-picene (14bα(H) & 14bβ(H)).	1,2	see Chaffee and Johns. (1983)	VI
B	2,2,4a,9-tetramethyl-1,2,3,4,4a,14b-hexahydropicenea.	1	see Chaffee and Johns. (1983)	VII
C	7-methyl-3'-ethyl-1,2-cyclopentenochrysene.	1-3	see Chaffee and Johns. (1983)	VIII
D	1,2,9-trimethyl-1,2,3,4-tetrahydropicene.	2	see Wakeham et al. (1980)	IX
E	2,2,9-trimethyl-1,2,3,4-tetrahydropicene.	2	see Wakeham et al. (1980)	X
F	Isomer of G and H of unconfirmed structure.	2	see Wakeham et al. (1980)	XI
G	2,8-dimethyl-1'-(2-propyl)-1,2-cyclopenteno-(1,2,3,4-tetrahydrophenanthrene).	4	see Laflamme & Hites (1979)	XII
H	3,3,7,12a-tetramethyl-1,2,3,4,4a,11,12,12a-octahydro-chrysene.	2,5	see Wakeham et al. (1980)	XIII

I	24-ethyl-4-methyl-19-norcho-lesta-1,3,5 (10)-triene.	XV	6	Hussler et al. 1981.
J	14α(H)-24-ethyl-1 (10→6)-abeo-cholsta-5,7,9-triene.	XVI	7	Hussler and Albrecht, 1983.
K	unknown			
L	unknown			

a May be 1,2,4a,9-tetramethyl (ef. ref. 1)

References: 9, 13,15, 16, 18, 29, 32.

* m/z 158(100%); 145(90%); 131(10%); 172(11.1%); 361(8.3%; M^+ −15); 376(5.7%; M^+)

¶ m/z 145(100%); 158(80%); 131(10%); 172(6.1%); 361(7.8%; M^+ −15); 376(5.7%, M^+).

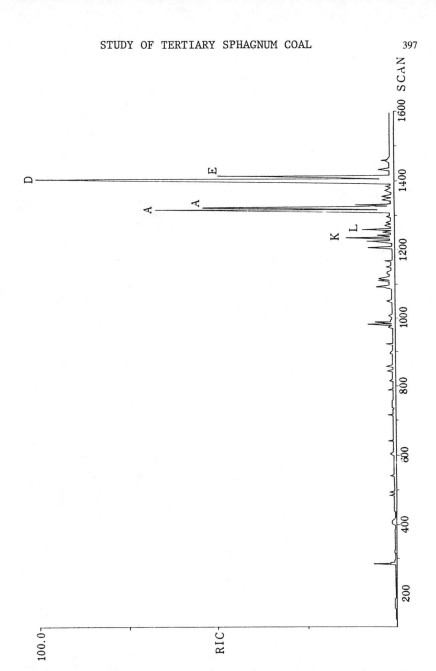

Fig. 6: Total ion reconstruction trace of the aromatic
hydrocarbon fraction of sample JS-1. Compound
assignments for peaks A–L are given in Table 2.

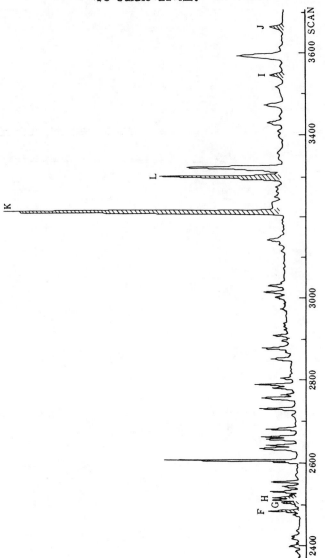

Fig. 7: Total ion reconstruction trace of the aromatic
        hydrocarbon fraction of sample JS-0 after separation
        of the tri- and tetraaromatic hydrocarbons seen in
        Fig. 5.   Structural assignments for components F-L
        are given in Table 2.

Many aspects of the biomarker distributions in the sphagnum coal are consistent with inputs of organic matter from higher plants. These features include: the dominance of higher n-alkane homologues (11,28), the presence of pentacyclic triterpenoids and tricyclic diterpenoids attributable to higher plant origins (4) and their aromatised derivatives, and the high ratio of hopanes to steranes (14). In the aliphatic hydrocarbons the abundance of higher, dominantly odd-numbered, n-alkanes maximizing at C_{31} (Fig. 3b) has been recognized previously in Sphagnum peats (10, 35) although such a distribution is also characteristic of higher plant waxes (e.g., 11). Olean-12-ene (Peak 2 in Fig. 4) is a common component of lignites and brown coals (15, 6, 36), often occurring, although not in this instance, with its more stable $\Delta^{13(18)}$ isomer. Taraxer-14-ene (Peak 1 in Fig. 4) is present in Tertiary and Quaternary deep-sea sediments from the Japan Trench (Brassell, unpublished confirmation of unknown triterpene in 3), but it has been observed in greatest abundance in a peat layer rich in Sphagnum remains from the English Lake District (35). The presence of both of these pentacyclic triterpenes in the sphagnum brown coal is therefore consistent with previous results, yet it is remarkable that they are not accompanied by hopenoid triterpenes (e.g., 35, 6, 36). The relatively minor amounts of tricyclic diterpenoid alkanes (Fig. 3a) and aromatics contrasts with the abundance of such components in resinous coals (25, 30) with major contributions from gymnosperms rather than angiosperms.

Sphagnum brown coal exhibits an unusual aromatic hydrocarbon composition featuring predominant pentacyclic tri- and tetraaromatic compounds and tetracyclic diaromatic compounds. Comparison of mass spectra of these tri- and tetra-

aromatic compounds with those of Victorian brown coal suggests that the structural skeletons of components A, B, D
and E belong to the oleanane and ursane families whereas
component C is a hopanoid derivative (9). The structures
of components F, G and H suggest pentacyclic triterpenoid
precursors (34), with either a six-membered E-ring (e.g.,
-amyrin, i.e., olean-12-en-3β-01, XVII) or a five membered
E-ring (e.g., lupeol, i.e., lup-20(29)-en-3β-01, XVIII; 19).

A range of aromatic hydrocarbons derived from pentacyclic triterpenoids similar to that of the sphagnum brown
coal has been recognized previously in a Miocene brown coal
(9) and in various recent sediments (19, 34, 3, 33, 13, 14,
31). In all cases, these components are attributed to
higher plants, specifically angiosperms. Their presence in
the sphagnum brown coal is therefore consistent with the
presumed major sources of its organic constituents.

The only monoaromatic steroid hydrocarbons detected
in the sphagnum brown coal were C_{29} components. A dominance
of C_{29} steroids is a typical feature of brown coals and lignites (e.g., 15, 6, 36); also, C_{29} monoaromatic anthraster-
oids (Peak J in Fig. 7) are present in Soviet brown coals,
although they were misassigned as 1-methyl ring A monoaromatics (24). In the sphagnum brown coal only the 14 (H)-
anthrasteroid isomer was detected, illustrating the immaturity of the sample, since such steroids readily isomerise to
their more thermally stable 14 (H) counterparts (16, 5).

Among the biological marker compounds that attest to
the immaturity of the samples are the 17 (H), 21 (H)-hopanes,
and various alkenes: olean-12-3n3, tarax-14-ene, and diterpenes (6). Such immaturity is consistent with the vitrinite
reflectance data. The occurrence of tri- and tetraaromatic
triterpenoids in such immature coal provides further evi-

dence for the aromatization of triterpenoids at a relatively early stage of diagenesis (9).

It is also relevant to note that few of the partially aromatized pentacyclic triterpenoids in the sphagnum brown coal have been recognized in mature sediments, bituminous coals or petroleums. Those that can survive at higher maturity levels are the more aromatized components, such as alkylated tetrahydropicenes (8, 32).

CONCLUSIONS

Recently, an unusual type of coal, sphagnum brown coal, was discovered in Yunan Province, China. It is composed of a significant proportion of sphagnum-textinite (72-85%) and less abundant sphagnum-fusinite. Its wax content is high (13-15%) and its maturity is low ($\bar{R}_{ran}$ 0.24%). Sphagnum brown coal is considered to be a type II kerogen with a good petroleum source potential. It is formed in an upland, nutrition-poor, swamp environment.

The high content of montan wax in sphagnum coal may be due to the high content of organic matter, within which large amounts of higher n-alkanes, with marked odd-over-even predominance are present. Terrigenous components are represented by the higher n-alkane homologues, plus higher plant triterpenoids, their aromatised derivatives, and the minor amounts of tricyclic diterpenoids and sesquiterpenoids. In addition, bacterial activity can be inferred from the occurence of hopenoids. Overall, the distributions of aliphatic and aromatic hydrocarbons show many characteristics that have been recognized previously in marine and lake sediments, sphagnum-rich peats, brown coals and lignites. Indeed, none of the biomarker features of the sphagnum brown coal appear to be specifically linked to, and therefore diagnostic of, inputs

from Sphagnum, although many are consistent with such contributions.

Although the brown coal is of low maturity (soft lignite
stage), it contains several, structurally-related, pentacyclic and tetracyclic angularly condensed aromatic hydrocarbons which can be inferred to be diagenetic products of naturally-occurring triterpenoid precursors (9). The presence
of such tri- and tetraaromatic triterpenoids demonstrates
that diagenetic aromatization of triterpenoids occurs at a
very early diagenetic stage in coals.

ACKNOWLEDGEMENTS

The authors are indebted to Mr. Zhou Yiping for the coal
samples and related geological data, to Dr. E. Evans and Mr.
B. Meloy for helpful discussions, and to Miss Li Hua for part
of the analytical work. This work was made possible by the
support of the Royal Society and the Academia Sinica. Partial
financial support was also provided by UNDP and UNESCO (CPR/
84/005).

REFERENCES

1. H. AGETA AND Y. ARAI,   Fern constituents: pentacyclic
    triterpenoids isolated from Polypodium niponicum and P.
    formosanum, Phytochem., 22, 1801-1808, (1983).

2. S. C. BRASSELL, The Lipids of deep sea sediments: their
    origin and fate in the Japan Trench, Ph.D. Thesis,
    University of Bristol, (1980).

3. S. C. BRASSELL, P. A. COMET, G. EGLINTON, P. J. ISAACSON,
    J. MCEVOY, J. R. MAXWELL, I. D. THOMSON, P. J. C. TIB-
    BETTS AND J. K. VOLKMAN, Preliminary lipid analyses of
    Sections 440A-7-6, 440B-3-5, 440B-8-4, 440B-68-2 and
    436-11-4: Legs 56 and 57, Deep Sea Drilling Project,
    Initial Reports of the Deep Sea Drilling Project 56/57,

Edited by Scientific Party (U.S. Government Printing Office, Washington, D.C., 1980), 1367-1391.

4.   S. C. BRASSELL, G. EGLINTON and J. R. MAXWELL, The geochemistry of terpenoids and steroids, Biochem. Soc. Trans., 11, 575-586, (1983).

5.   S. C. BRASSELL, J. MCEVOY, C. F. HOFFMANN, N. A. LAMB, T. M. PEAKMAN and J. R. MAXWELL, Isomerisation, rearrangement and aromatisation of steroids in distinguishing early stages of diagenesis, Advances in Organic Geochemistry, Edited by P. A. Schenck, J. W. de Leevw and G. W. M. Lijmbach, Org. Geochem, 6, 11-23, (1983).

6.   S. C. BRASSELL, G. EGLINTON and FU JIAMO, Biological marker compounds as indicators of the depositional history of the Maoming oil shale, Advances in Organic Geochemistry 1985, Edited by D. Leythaeuser and J. Rullkotter (Pergamon, Oxford) Org. Geochem., 12, 927-941, (1986).

7.   S. C. BRASSELL, G. EGLINTON and V. J. Howell, Paleoenvironmental assessment for marine organic-rich sediments using molecular organic geochemistry, Marine Petroleum Source Rocks, edited by J. Brooks and A. J. Fleet, (Blackwells, Oxford, 1987).

8.   W. CARRUTHERS and D. A. M. WATKINS, Identification of 1,2,3,4-tetrahydro-2,2,9-trimethylpicene in an American crude oil, Chem and Ind., 1433-1435, (1963).

9.   A. L. CHAFFEE and R. B. JOHNS, Polycyclic aromatic hydrocarbons in Australian coals. I. Angularly fused pentacyclic tri- and tetraaromatic compounds of Victorian brown coal, Geochim. Cosmochim. Acta, 47, 2141-2155, (1983).

10. P. A. CRANWELL, Chain-length distribution of n-alkanes from lake sediments in relation to post-glacial environmental change. Freshwater Biol., 3, 259-265, (1973).

11. G. EGLINTON and R. J. HAMILTON, Leaf epicuticular waxes, Science, 156, 1322-1335, (1967).

12. FU JIAMO, SHENG GUOYING, CHEN DEYU, LIU DEHAN, S. C. BRASSELL, A. GOWAR and G. EGLINTON, Geochemical characteristics of sphagnum brown coal -- A possible oil-generating precursor, Geochimica, 1, 1-9, (1987).

13. P. GARRIGUES, A. SAPTORAHARDJO, C. GONZALEZ, P. WEHRUNG,
    P. ALBRECHT, A. SALIOT AND M. EWALD, Biogeochemical aro-
    matic markers in the sediments from Mahakam Delta (Indo-
    nesia), Advances in Organic Geochemistry 1985, Edited
    by D. Leythaeuser and J. Rullköttes, Org. Geochem., 10
    959-964, (1986).

14. A. C. GREINER, C. SPYCKERELLE AND P. ALBRECHT, Aromatic
    hydrocarbons from geological sources. I. New naturally
    occurring phenanthrene and Chrysene derivatives, Tetra-
    hedron, 32, 257-260, (1976).

15. C. F. HOFFMAN, A. S. MACKENZIE, C. A. LEWIS, J. R. MAX-
    WELL, J. L. OUDIN, B. DURAND AND M. VANDENBROUCKE, A
    biological marker study of coals, shales and oils from
    the Mahakam delta, Kalimantan, Indonesia, Chem. Geol.,
    42, 1-23, (1984).

16. G. HUSSLER AND P. ALBRECHT, $C_{27} - C_{29}$ monoaromatic anthra-
    steroid hydrocarbons in Cretaceous black shales, Nature,
    304, 262-263, (1983).

17. G. HUSSLER, B. CHAPPE, P. WEHRUNG AND P. ALBRECHT, Iden-
    tification of $C_{27} - C_{29}$ ring A monoaromatic steroids in
    Cretaceous black shales, Nature, 294, 556-558, (1981).

18. J. KARLINER AND C. DJERASSI, Terpenoids. LVII. Mass Spec-
    tral and nuclear magnetic resonance studies of pentacy-
    clic triterpene hydrocarbons, J. Org. Chem., 31, 1945-
    1956, (1966).

19. R. E. LAFLAMME AND R. A. HITES, Tetra- and pentacyclic
    naturally occurring aromatic hydrocarbons in recent sedi-
    ments, Geochim. Cosmochim Acta, 43, 1687-1691, (1979).

20. A. LIVSEY, A. G. DOUGLAS AND J. CONNAN, Diterpenoid hy-
    drocarbons in sediments from an offahore (Labrador) well,
    Org. Geochem., 6, 73-82, (1984).

21. A. S. MACKENZIE, R. L. PATIENCE, J. R. MAXWELL, M. VAN-
    DENBROUCKE AND B. DURAND, Molecular parameters of matur-
    ation in the Toarcian shales, Paris Basin, France, I.
    Changes in the configurations of acyclic isoprenoid al-
    kanes, steranes and triterpanes, Geochim. Cosmochim.
    Acta, 44, 1709-1721, (1980).

22. A. S. MACKENZIE, C. F. HOFFMANN AND J. R. MAXWELL, Mo-
     lecular parameters of maturation in the Torcian shales,
     Paris Basin, France III. Changes in aromatic steroid hy-
     drocarbons, Geochim. Cosmochim. Acta, 45, 1345-1355, (1981).

23. J. M. MOLDOWAN, W. K. SEIFERT AND E. J. GALLEGOS, Rela-
     tionship between petroleum composition and depositional
     environment of petroleum source rocks, Amer. Assoc. Pe-
     trol. Geol. Bull., 69, 1255-1268, (1985).

24. A. A. PETROV, N. S. VOROBIEVA AND ZEMSKOVA, Z. K., Ster-
     enes and triterpenes in brown coals, Org. Geochem., 8,
     269-273, (1985).

25. R. A. NOBLE, R. ALEXANDER, R. I. KAGI AND J. KNOX, Tetra-
     cyclic diterpenoid hydrocarbons in some Australian coals,
     sediments and crude oils, Geochim. Cosmochim. Acta, 49,
     2141-2147, (1985).

26. W. K. SEIFERT AND J. M. MOLDOWAN, Paleoreconstruction
     by biological markers, Geochim. Cosmochim. Acta, 45,
     783-794, (1981).

27. SHENG GUOYING, FU JIAMO, ZHOU ZHONGYI AND SHEN RULANG,
     Benzohopanes -- A novel family of biomarker compounds
     detected in Jurassic sedimentary rocks, Geochimica, Sin-
     ica, 75-79, (1985).

28. B. R. T. SIMONEIT, The organic chemistry of marine sedi-
     ments, Chemical Oceanography, Vol. 7, Edited by J. P.
     Riley and R. Chester, 233-311, (1978).

29. B. R. T. SIMONEIT AND M. A. MAZUREK, Organic matter of
     the troposphere, II. Natural background of biogenic
     lipid matter in aerosols over the rural Western United
     States, Atmospheric Environment, 16, 2139-2159, (1982).

30. B. R. T. SIMONEIT, J. O. GRIMALT, T. G. WANT, R. E. COX,
     P. G. HATCHER AND A. NISSENBAUM, Cyclic terpenoids of
     contemporary resinous plant detritus and of fossil woods,
     ambers and coals, Advances in Organic Geochemistry 1985,
     edited by D. Leythaeuser and J. Rullkötter, Org. Geochem.,
     10, 877-889, (1986).

31. C. SPYCKERELLE, A. C. GREINER, P. ALBRECHT AND G. OURIS-
     SON, Aromatic hydrocarbons from geological sources. IV.
     An octahydrochrysene, derived from triterpenes, in oil

shale: 3,3,7,13-tetramethyl-1,2,3,4,11,12,13,14-
octahydrochrysene, J. Chem. Res., (M), 3801-3828; (S),
332-333, (1977).

32. M. STREIBL AND V. HEROUT, Terpenoids- especially oxy-
genated mono-, sesqui-, di- and triterpenes, Organic
Geochemistry: Methods and Results, edited by G. Eglinton and M. T. J. Murphy (Springer, Berlin), 401-424,
(1969).

33. Y. L. TAN AND M. HEIT, Biogenic and abiogenic polynuclear
aromatic hydrocarbons in sediments from two remote
Adirondack Lakes, Geochim. Cosmochim. Acta, 45, 2267-
2279, (1981).

34. S. G. WAKEHAM, C. SCHAFFNER AND W. GIGER, Polycyclic
qromatic hydrocarbons in recent sediments, II. Compounds derived from biogenic precursors during early
diagenesis, Geochim. Cosmochim. Acta, 22, 415-429, (1980).

35. A. M. K. WARDROPER, Aspects of the geochemistry of polycyclic isoprenoids, Ph. D. Thesis, University of Bristol,
(1979).

36. E. WINKLER, Organic geochemical investigations of brown
coal lithotypes. A contribution to facies analysis of
seam banding in the Helmstedt deposit, Advances in Organic Geochemistry 1985, edited by D. Leythaeuser and J.
Rullkötter, Org. Geochem., 10, 617-624, (1986).

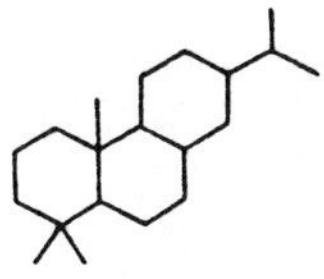

I

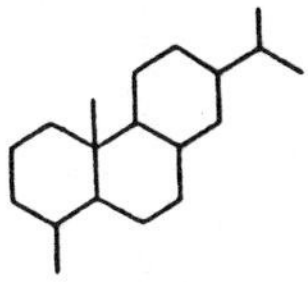

II

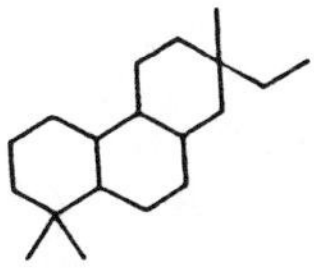

III

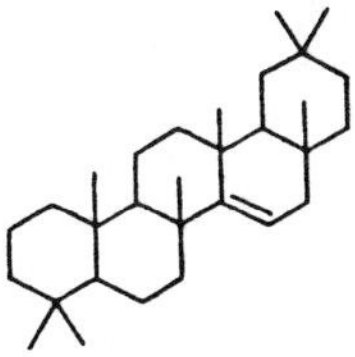

IV

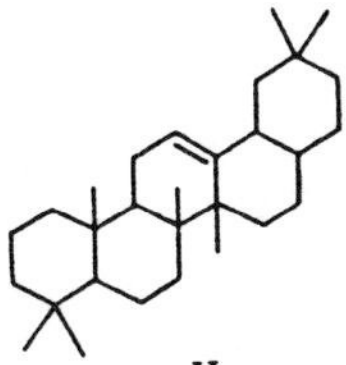

V

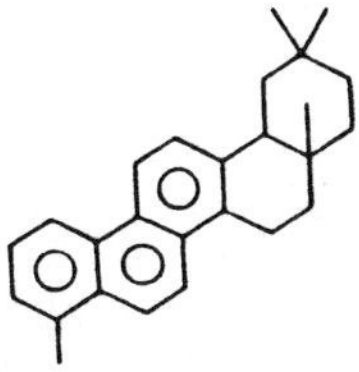

VI

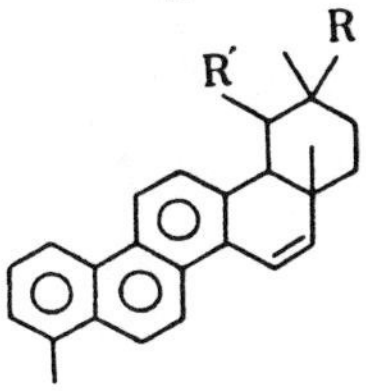

:VII

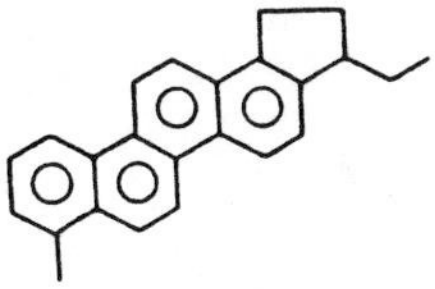

VIII

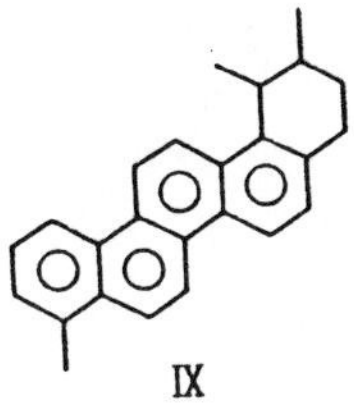

IX

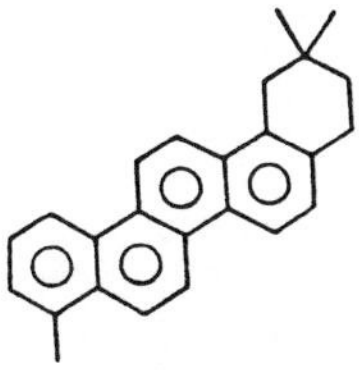

X

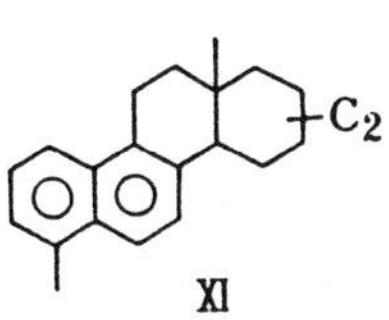

XI

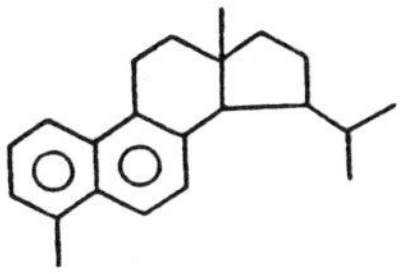

XII

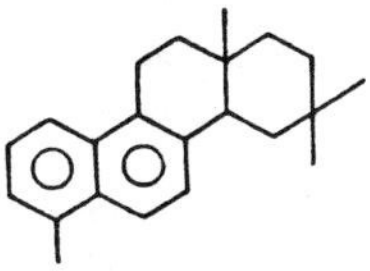

XIII

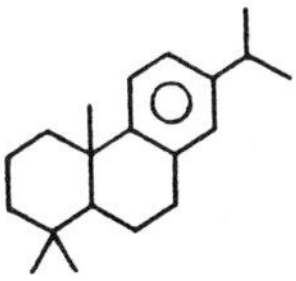

XIV

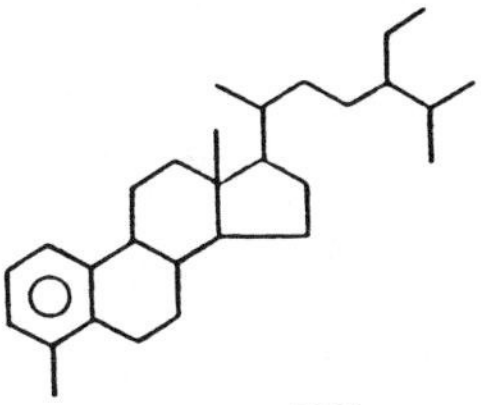

XV

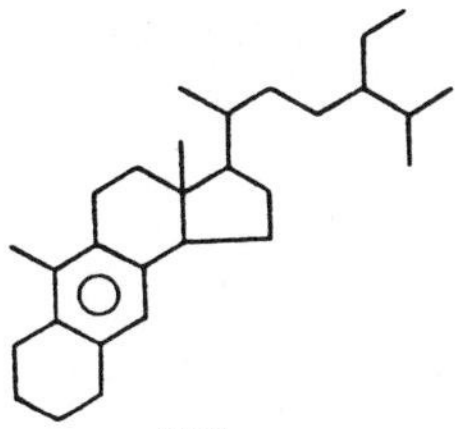

XVI

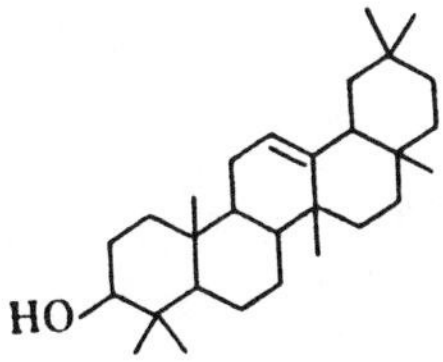

XVII

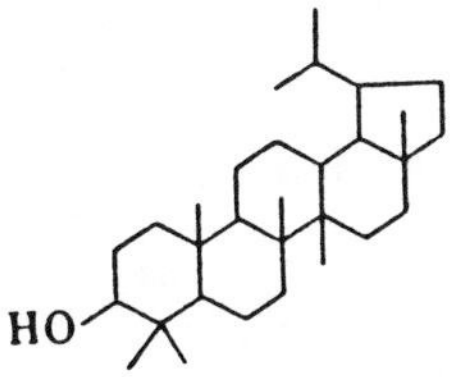

XVIII

Appendix

APPLICATION OF BIOMARKERS FOR CHARACTERIZATION OF NATURAL WATERS: BLACK TRONA WATER

YU WANG[*], LIEN-SEN WANG[**] AND TEH FU YEN
Environmental and Civil Engineering
University of Southern California
Los Angeles, CA 90089-0231

The organics in the black trona water of southwestern Wyoming are studied in their geochemical biomarker nature pertaining to aromatic hydrocarbons, alkanes and acyclic isoprenoids, pentacyclic triterpenoids, steranes and tri- and tetra-cyclic terpanes. Results obtained in the aspects of source materials, maturation, biodegradation, and migration or conditions in the paleoenvironments are helpful in reconstructing a plausible hypothesis of the origin and fate of the massive organics and their diversities of the black trona water known today.

* Present address: Department of Environmental Engineering, East China University of Chemical Technology, Shanghai, PRC.

**Present address: Department of Environmental Chemistry, Nanjing University, Nanjing, PRC.

1.   INTRODUCTION

Major interaction between hydrosphere and lithosphere can
take place in two ways; either by erosion, weathering and
dissolution of lithosphere to hydrosphere, or by precipitation, deposition and sedimentation of hydrosphere to
lithosphere. Various stages of either process will create
bodies of water in given geological confinement with high
mineral and organic contents, although the second type of
interaction is of special importance, since the lithofi-
cation process may involve biosphere participation. The
mineral water or brine derived may contain a number of
biomarkers beyond the dissolved major organic species.

Ideally, natural water candidates used for investigation should include the following requirements:

- (1)   Stagnant body of water under some sort of
        equilibrium (e.g., aquifer.)
- (2)   No anthropogenic input.
- (3)   Fixed values of pH and E_h.
- (4)   Adequate level of dissolved organics.
- (5)   High loading of inorganic salts with buffer.
- (6)   Presence of surfactant.
- (7)   No profuse bacterial growth.

There is an artesian aquifer of several hundred
square miles at depths often less than 1000 ft. in southwestern Wyoming (Fig. 1). Geologically, this aquifer lies
in the northern Green River Basin and belongs to the New
Fork Tongue Member of the Wasatch Formation. Immediately
above the New Fork Tongue Member is the well-known Wilkins
Peak Member which contains the richest oil shale of the
Green River Formation (Fig. 2).

Strangely, some of the wells drilled in this region

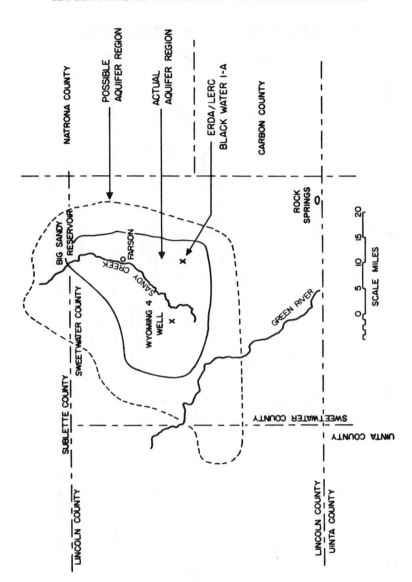

FIGURE 1 Location of Black Water and artesian aquifer in northern Green River Basin.

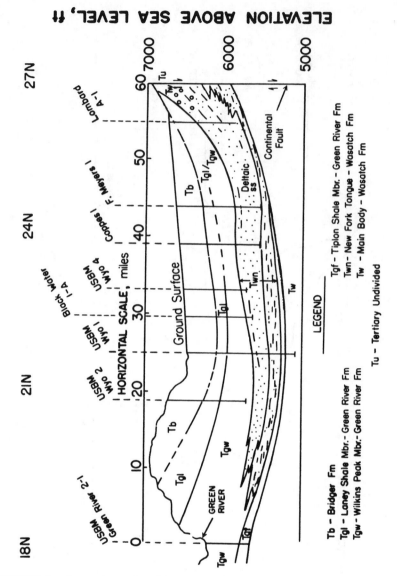

FIGURE 2   Stratigraphy diagram of sediments in Eocene-
Paleocene time of Green River Basin.   Locations of wells
drilled from south to north are shown.

encountered an unusual black trona water which contains
organic acids dissolved in sodium carbonate brine. This
water is under pressure similar to that found in artesian
aquifer. All the wells indicated in Fig. 2 yield brine
waters, yet the water quality differs. Only two wells out
of eight, USBM Wyoming 4 and Black Water 1-A, give black
trona water.(see Fig. 2) (1-3). Apparently, the encounter
of black trona water is limited to those wells which were
just below the Wilkins Peak Member but within the New Fork
Tongue. Drilling either above or below the New Fork
Tongue has not yielded any black trona water. Also it
seems that drilling has to be within a 26-foot zone located below the top of the Wilkins Peak Member but slightly
penetrating the fluvial and deltaic wedge in the New Fork
Tongue.. In this case, only mineral-laden brine water,
absent of polymeric organic acids of dark color, is encountered in the middle or near the bottom of the New Fork
Tongue.

Evidence in recent findings (3,4) has indicated that
the black trona water is most likely ancient lake water
trapped in sedimentary depressions on the lake bottom,
caused by recessions and advances of the lake. As the result of the algal biomass turned into oil shale during the
Eocene epoch, the eutrophication process of Lake Gosiute
starting from ultraoligotrophic to oligotrophic, to mesotrophic, to eutrophic, to hypereutrophic, and eventually to
today's Green River landscape had left the surviving remnants of water pods and lenses. These preserved black
waters may represent the geochemical environment of an
Eocene Lake.

Closely associated with the black trona water was, of
course, the deposit of Green River Oil Shale. The organic

portion of the shale can either be kerogen (insoluble in organic solvents) or bitumen (soluble in most organic solvents). These fossil remains are believed to be formed from the cyanophycèa, or blue-green algae, which grow in Lake Gosiute. Therefore, correlation is important to see if the preserved organic matter in trona water is similar to that preserved in oil shale.

2. EXPERIMENTAL

The black trona water used in this work was from ERDA/LERC Black water 1-A well (Fig. 1 and 2). The sample as received was in a polypropylene container and immediately refrigerated. It was a dark brown, viscous, syrup-like liquid with a pH value of 10.58. A TOC determination of this water was measured as 12,360 mg/L by means of Dohman model 60 total organic carbon analyzer.

This water was slowly neutralized with hydrochloric acid to pH ≈ 7 and extracted with dichloromethane (2:1 v/v) in a liquid-liquid extractor. After drying with anhydrous sodium sulfate, the dichloromethane extract was directly used for capillary GC measurement. After evaporation of the solvent, the residue was used for IR determination.

Hewlett Packard model 5880A GC equipped with OV-101 fused silica capillary column and a FID detector was used for investigation. The injection and detector temperature was 250°C and programmed at 10°C/min from 30–280°C with the gas flow at 3 ml/min as carrier gas. IR is from Bechman infrared spectrophotometer Acculab 6.

GC-MS was carried out with Finnegan 4000 coupled to Incos Data System 2300. A 30 m long with 0.25 mm I.D. DW5 capillary column was used. Temperature program was as follows: 35°C, 6 min; 35–280°C, 4°C/min; 280–310°C,

2°C/min.

3. RESULTS AND DISCUSSION

Aromatic hydrocarbons

Two separate methods have been used for the identification of individual aromatic hydrocarbon, one is by
coinjection with known standard hydrocarbons, the other is
by employing mass spectrometry. For the latter, we have
specifically monitored m/z = 91, 128, 134, 152, 170, 178,
202, 252, 278 . . . which is equivalent to the characteristic ions of the given aromatic class.

A number of polynuclear hydrocarbons are present in
black trona water, and their concentrations are significantly high. For example, the total amount of hydrocarbons
listed in Table 1 is almost 0.4 g/L out of the total organic carbon of 12.36 g/L in the water. The presence of a
large amount of condensed aromatic hydrocarbon species is
perhaps surprising, since the Green River oil shale kerogen is quite low in aromaticity (e.g. $f^{a} \leq 0.20$) (5,6).
However, Branthaver and Barden (7) investigated the polymeric acid portion of the black trona water, which consists
of 40-45% of the TOC's, and found to contain a largely
condensed aromatic system of 3-5 rings. They studied this
portion by both ^{13}C-NMR and fluorescence spectroscopy. It
is possible that the non-polymeric portion (low molecular
weight soluble fraction) may contain molecules of PNAs as
illustrated in Table 1.

It is important to note that Gallegos (11) has
identified a number of mono- and diaromatics in Green
River Oil Shale bitumen, particularly substituted
phenylcyclohexyl alkanes, alkylbenzenes, monoaromatized
terpanes or steranes by use of fragment ion at m/z 119

TABLE 1   PNAs IN BLACK WATER BY GC/MS

Compound	Approximate Amount (dg/L)
propyl-benzene	0.030
1-propenyl-benzene	0.025
1,2,3,4-tetramethylbenzene	0.269
naphthalene	0.084
2,6-dimethylnaphthalene	0.807
acenaphthene	-
fluorene	-
2,3,5-trimethylnaphthalene	0.170
dibenzothiophene	0.163
phenanthrene	0.249
anthracene	0.153
chrysene	-
pyrene	0.322
1,2-benzoanthracene	-
benzo(e)pyrene	0.356
benzo(a)pyrene	0.292
perylene	0.388

(Fig. 3 a to f). This large amount (31%) of the monoaromatics in Green River oil shale bitumen is associated with
substituted tetralins (12). Probably 1,2,3,4-tetramethyl-
benzene (prehnitol) (Fig. 3 i) in Table 1 is due to mono-
aromatization (B ring) of isocopalane type structure (see
Fig. 9). Similarly, 2,6 and 2,3,5-methyl substituted naphthalenes (Table 1) may originate from the cleavage of BC-
ring diaromatic steroids (14). Doubtlessly both phenanthrene
and chrysene (Table 1) are in accordance with the orienta-
tion of tri- and tetra cyclic terpenes. There is no 4,5-
substituted phenanthrene suggesting an origin of aromatics
from isoprenoids.

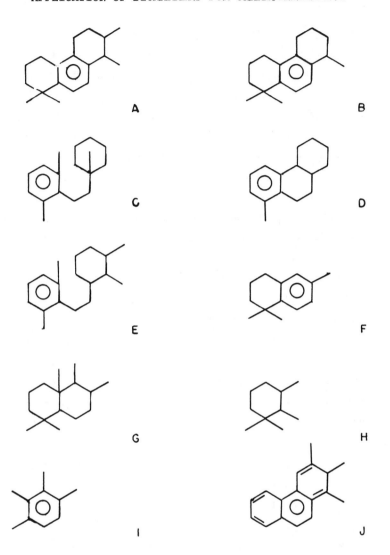

FIGURE 3   Monoring Aromatics and Cycloparaffins in Green
River Oil Shale.   a to f are among compounds
studied in Ref. g is methyl-substituted decalin,
drismane is from Bendoraitis, Ref. 33 h is tetra-
substituted cyclohexane from Anders and Robinson,
Ref. 26 i is prehnitol j is 1,2,3-trisubstituted
phenanthrene.

Alkanes and Acyclic Isoprenoids

According to conventional methods for the determination of n-alkanes at m/z 99, a distribution pattern is plotted with the calibrated standards (Fig. 4). Occasionally even carbon numbers are pronounced (C_{20}, C_{30}, etc.) The overall CPI calculated is 1.15 with C_{17} at peak ranging from C_{11} to C_{33}. The qualitative evaluation is listed in Table 2. Basically, the distribution pattern of Fig. 4 resembles that of the Green River Oil Shale bitumen (15). Peaks at C_{15} and C_{17} are from algal contribution, whereas peaks at C_{27} and C_{29} are due to fungi and other higher plants (16).

Fig. 5 indicates acyclic isoprenoids which have been studied. IC_{18}, IC_{19} and IC_{20} are the most pronounced where there is no IC_{17} and IC_{22}. The Pr/Ph ratio is about 0.5 suggesting that, under the strong alkaline condition of the brine, it is anoxic condition in deposition. The even predominance of n-alkanes (e.g. C_{20}, C_{30} in Fig. 4) is always associated with the predominance of phytane over pristane (17). In the present case, apparently the pH of the bottom of Lake Gosiute is believed to have been between 10 and 4, a reducing environment having been in existence. These facts are in agreement with the conclusion of Smith et al. (4) and Smith (18).

Pentacyclic Triterpenoids

Ever since Henderson et al. (19) identified hopane in Green River Oil Shale bitumen, enormous progress has been made in the application of hopanes as biomarkers in various types of sedimentary rocks and fossil fuels. Due to the ubiquitous nature of hopanes in sediments, Ourisson et al. (20) believed that hopanoids are rigidifiers for cell wall membranes. Hopanes are known to be widely dis-

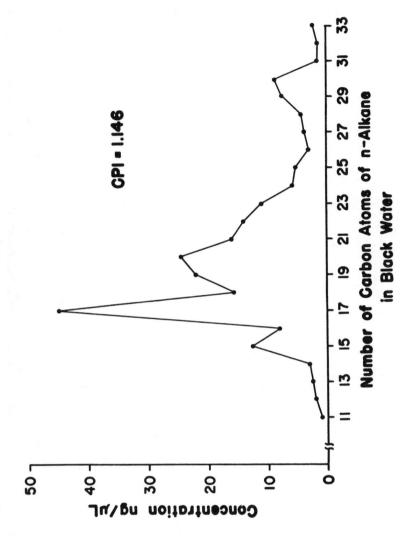

FIGURE 4   Distribution pattern plotted with the calibrated
standards according to conventional methods for
the determination on n-alkanes at m/z 99.

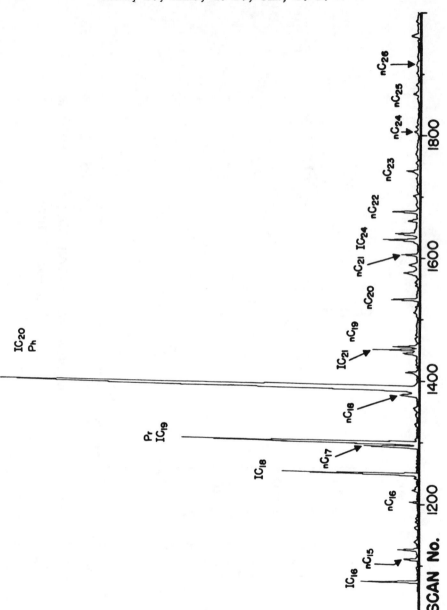

FIGURE 5   Spectrum of acyclic isoprenoids and n-alkanes.

TABLE 2   n-ALKANE IN BLACK WATER

C No.	Conc. (mg/L)
Standard	15.0
11	0.98
12	1.88
13	2.25
14	2.81
15	13.13
16	8.44
17	45.0
18	15.38
19	22.50
20	24.38
21	15.94
22	14.06
23	11.63
24	5.63
25	5.63
26	2.81
27	3.75
28	3.75
29	7.50
30	8.44
31	1.88
32	1.88
33	2.25

tributed among cyanobacteria (blue-green algae), bacteria and ferns (plants).

Table 3 lists the triterpanes identified in black trona water. Fig. 6 is the spectral records of these triterpanes. They range from C_{27} to C_{31} with C_{28} missing as the case in most sediments (including Green River Oil Shale bitumen). For black trona water, Tm > Ts; Tm is in the hopane series, yet the origin of Ts is questionable, since there is no known alternative source for Ts. Tm/Ts can serve as a maturity indicator if another lens or trap of Eocene aquifer is located.

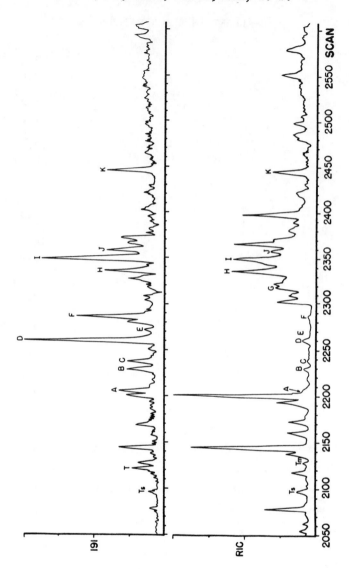

FIGURE 6   Spectral records of the triterpanes identified
in black trona water listed in Table 3.

Table 3   Triterpane in Black Water

Compound	Name	M.W.	Identification
$C_{27}H_{46}$	18α(H)-22,29,30-trisnorhopane	370	Ts
$C_{27}H_{46}$	17α(H)-22,29,30-trisnorneohopane	370	Tm
$C_{29}H_{50}$	17α(H),21β(H)-30-norhopane	398	A
$C_{30}H_{50}$	hop-17(21)-ene	410	B
$C_{29}H_{50}$	17β(H),21α(H)-normoretane	398	C
$C_{30}H_{52}$	17α(H),21β(H)-hopane	412	D
$C_{30}H_{52}$	unknown C_{30} triterpane	412	E
$C_{30}H_{52}$	17β(H),21α(H)-moretane	412	F
$C_{31}H_{54}$	22S-17α(H),21β(H)-30 homohopane	426	G
$C_{31}H_{54}$	22R -17α(H),21β(H)-30 homohopane	426	H
$C_{30}H_{52}$	gammacerane	412	I
$C_{30}H_{52}$	17β(H),21β(H)-hopane	412	J
$C_{30}H_{54}$	17β(H),21α(H)-homomoretane	426	J
$C_{31}H_{54}$	17β(H),21β(H)-30 homohopane	426	K

Stereoisomerization at C_{17} and C_{21} gives the value of 17α(H), 21β(H)/17β(H), 21β(H)=2.2. The biological input of 17β(H), 21β(H) configuration has matured into 17α(H), 21β(H) configuration under bury. This fact is again supported by the chiral center at C_{22}. The calculated value of 22S/22R is 5.6 which corresponds to a maturation parameter of 22R/(22R + 22S) as 0.15. Apparently the biomass from cyanobacteria has undergone maturation in some extent under sealed conditions during the close of the Eogene Lake. There is, however, no severe biodegradation due to the fact that there are no 25-norhopane observed. Perhaps this implies that the biomass and the caustic ingredients of the black trona water are simultaneously deposited.

The presence of gammacerane (I) in the water might indicate another source input, e.g. protozoan, since tetrahymanol is related to protozoan. Similarly, the presence of the unknown C_{30}-pentacyclic triterpane (E) (equivalent to compound No. 6 of Ref. 13) represents another terrestrial source.

Steranes

In general, the m/z 217 spectrum is quite simpler (Fig. 7), represented by fewer than 15 compounds. Clearly, this is an indication that this black trona water is non-marine (21) and there is at no time any intrusion of the Cannon-ball Sea. All steranes from C_{27} to C_{29} are found to contain both 5α(H) and 5β(H) isomers, and their ratios are between 7:1 and 8:1. Gallegos (22) reported that the ratio for the Green River Oil Shale bitumen is 3:1 for C_{27}, C_{28} and C_{29} steranes. It is possible that when the biological precursors were reduced to an isomeric mixture, e.g. cholesteral to cholestane (5α) and coprostane (5β), the trans or the 5α(H) configuration is the more stable of the two. Likely, different hydrogenation mechanisms would yield different ratios of 5α(H) to 5β(H) as the product.

Surprisingly, there is no configurational isomerized of any 14β(H) or 17β(H) products. Every sterane in Fig. 7 is 14α, 17α, in configuration. Since sterane precursors, e.g., sterols, require a flat configuration as an insert in membrane structure, the best configuration is 8β(H), 9α(H), 10β(CH$_3$), 13β(CH$_3$), 14α(H), 17α(H) (24). Whereas C-8 and C-9 exist as the most stable configuration, C-10 and C-13 cannot be changed by hydrogen exchange, thus 14α(H), 17α(H) are already the stable ones. Since in Fig. 7 the cholestane, ergostane and stigmastane are all with 14 α(H), 17 α(H)

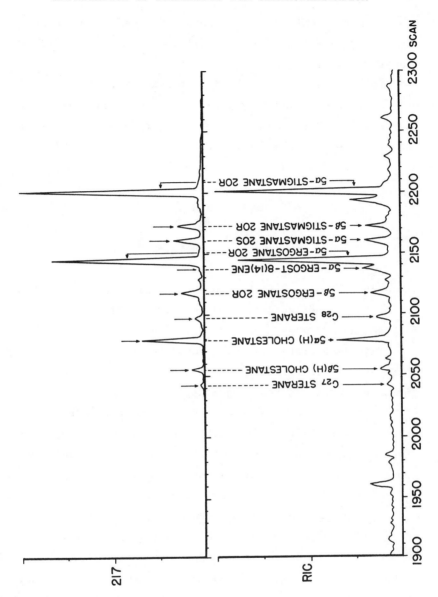

FIGURE 7   Distribution pattern of steranes (m/z at 217) with the compounds identified.

configuration, this indicates that the organics in black trona water are immature and still in the first stage of transformation.

To support this hypothesis, steranes so far identified only have 20R-configuration, excepting one. Mackenzie (23) has used 20S/(20R + 20S) as an indicator for maturity. In the organics from black trona water the low value indicated immaturity. This can be understood by the following:

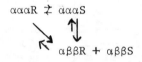

The carbon sequence of the above expression is 5, 14, 17, 20 for the isomerization. The only 20S in Fig. 7, 5α ergostane 20S has a configuration of ααα S.

This immaturation is also supported by the fact that there are no diasteranes found since any diacholestane is formed by catalytic rearrangement.

Speaking about source material, ergostane probably originated from algae (from dihydroergosterol) and stigmastane and cholestane, from other plants (stigmasterol and cholesterol). There is no 21-norcholestane found, probably suggesting fungi input is small (from fugisterol).

Tri- and Tetra-cyclic Terpanes

Moldowan et al. (25) found an homologous series of tricyclic terpanes ranging from $C_{19}H_{34}$ to $C_{45}H_{86}$ (Fig. 8 e, when C_{14} is H, the C_{19} as beginning series) to the near ubiquitously distributed to crude oil and organic-rich sediments. For example, more than 12 tricyclic terpanes have been identified in Green River Oil Shale bitumen (26, 27, 28). The m/z 191 spectrum in Fig. 9 indicates both the distribution

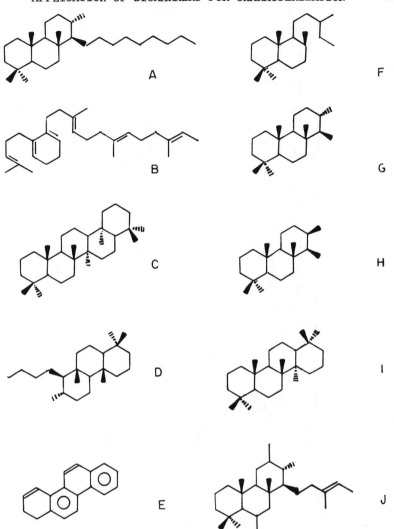

FIGURE 8   Tri- and tetracyclic terpanes and related compounds.
a, labden C_{20}; b, C_{20} Tricyclic diterpane, b can
be found through a; c, cheilanthatriol; d, cheilanthane found in Ref. 34   as 18,19-bisnor 13 β(H).
14α(H) cheilanthane; e, C_{30} tricyclic triterpane
(n=1), C_{40} tricyclic tetraterpane (n=3), when n=4
a C_{45} tricyclic terpane is obtained; f, isocopalane; g, an hexaisoprenoid; h, tetracyclic terpanes,
R = H, C_{24} 17,21-secohopane; i, gamacerane; j,
chrysene.

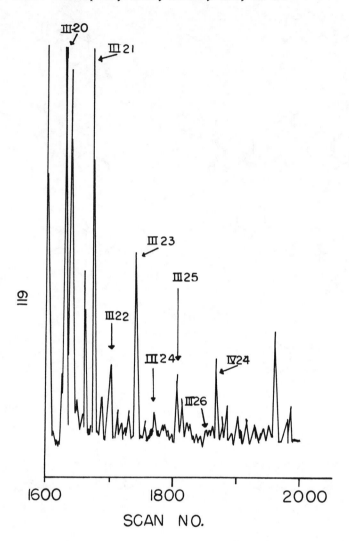

FIGURE 9   Tricyclic terpanes (with one tetracyclic terpane)
as determined by single ion monitoring at m/z 191.

of the tricyclic terpanes (III-series) from C_{20} to C_{26} and the tetracyclic terpanes (IV-series) of C_{24}. As judged by intensity, apparently both the C_{20} and C_{21} compounds take the bulk of the tricyclic terpanes. We will discuss this further.

The biological precursors for these tricyclic terpanes are not clear. The C_{30} compound (Fig 8 e) has been proposed to be derived from hexaprenal through six isoprene units (Fig. 8 g) (29). Ekweozor and Strausz (30) postulated that the C_{23} compound of this series (Fig. 8 d) is from sesterterpenoids such as cheilanthatriol (Fig. 8 c).

In a study of biomarkers for Green River Oil Shale bitumen, Gallegos found that the C_{20} and C_{21} compounds of the tricyclic terpanes account for 80% of the total (22). This fact corresponds to our findings of the organics in the black trona water. Therefore, the abundance of C_{20} and C_{21} terpanes must imply that original biological precursors in Lake Gosiute are derived by more than one mechanism, i. e., they can come from di-, sester-, and tri-terpanes. For example, a C_{20} labden (Fig. 8 a) may proceed by maturation to a C_{20} tricyclic diterpane (Fig. 8-b) with configuration of $13\beta(H)$, $14\alpha(H)$ identical to cheilanthane (Fig 8 d). Although an isomer, the $13\beta(H)$, $14\beta(H)$ has been found from Athabasca oil sand bitumen (30), so far the $13\beta(H)$, $14\alpha(H)$ isomer of the isocopalane configuration (Fig. 8 f) has not been reported. Degradation of hopanoid compounds to this series (Fig. 8 b and Fig. 8 e) is possible.

Reviewing the monoaromatics isolated by Gallegos (11) in Fig. 3 (a to f), their configurations are always consistant with the C_{20} and C_{21} tricyclic terpanes which are highest in amount (Fig. 8). Since we have a large amount of phenterthrene (Fig. 3 j) (Table 1) aromatization in the course of disproportiation must proceed with ease.

Similarly, a C_{24} tetracyclic terpane was found (Fig. 9). Actually, this is a 17,21-secohopane, C_{24} (31). Reviewing all possible hypotheses for this type of compound, the one which suggests that microbial ring scision of hopanoids at an early state of diagenesis followed later by geochemical reduction is more likely. Probably biodegradation of hopenoid hydrocarbons in Lake Gosiute during oligotrophic and mesotrophic stages and followed by the nonbiological reduction beginning at eutrophic stage and ending at biostratinomy and oryclocenosis to fossil assemblage. It was estimated that the existance and disappearance of Lake Gosiute and Lake Uinta in the tri-state area during Eocene time was approximately 4 to 5 million years.

4. CONCLUSION

We have applied the biomarker techniques for the study and characterization of a natural water system, the black trona water from southwestern Wyoming. This water represents the paleo environments between Tipton Shale Member and Laney Shale Member periods (Fig. 2). For Lake Gosiute, the Laney Shale Member of Green River Formation is the last stage at Late Eocene time.(32). Current study gives the understanding that the organic matter in Lake Gosiute water which escaped by evaporation was converted to the kerogen and bitumen of Green River Oil Shale. Possibly the sequences of the occurence and fate of black trona water are as follows:

1. During Middle Eocene time, Lake Gosiute and Lake Uinta must have been connected, and at Late Eocene time after the lake dried up, the four major oil shale producing basins were formed; the Green River Basin, Washakie Basin,

Uinta Basin and Piceance Creek Basin.

2. The organics in the black trona water are related to the organics in Green River Oil Shale.

3. Excess growth of fungi in the lake caused stratification of epilimnion and hypolimnion.

4. Hydrolysis of silicates furnished sodium and created a basic solution in the bottom of the lake.

5. Respiration exceeded photosynthesis; the consequence of this would result in the decomposition of organic matter with excess carbon dioxides.

6. Concentration of sodium carbonate would increase to cause the deposition of nahcolite ($NaHCO_3$), darsonite ($NaAl(OH)_2CO_3$), shortite ($Na_2CO_3 \cdot 2CaCO_3$) and trona ($NaCO_3 \cdot NaHCO_3 \cdot 2H_2O$).

7. The Rock spring uplift might have caused the split of Lake Gosiute, especially during the hypereutrophic stage. Black trona water appeared as a remnant of the trapped lense at that time.

8. During eutrophic and posteutrophic stages, most lipids would be saponified, e.g., the sodium salts of fatty acids (8,9), of humic acids (7,10), of the deoxyphylloerythrin (C_{33} acids) and its homologeous acids (4).

9. During the period of the formation of isolated lenses at Green River basin until the present, reducing environments causing hydrogenation and dehydrogenation, the latter is caused by labile hydrogen shift, for the formation of many aromatic systems.

10. Creation of today's black trona water is probably due to thrust of New Fork Tongue from Rock Spring uplift to the West (Fig. 2) and the entrapment of the extensive growth of fungi in the Wilkins Peak Member time in an arheic region.

5. ACKNOWLEDGEMENTS

Support of Petroleum Research Fund Grant through the American Chemical Society of 16319-AC512 and DOE Contract No. A503-76-GU10017 is greatly appreciated. Both Y. W. and L.-S. W. thank their respective organizations in PRC for providing them the opportunity to come to U.S.C. as visiting scholars. The authors are especially thankful for the black trona water samples obtained through Dr. J. W. Smith of the Department of Energy, Laramie, Wyoming.

REFERENCES

1. Dana, G. F. and Smith, J. W., 1973, Black Trona Water, Green River Basin, 25th Field Conference—1973. Wyoming Geological Association Guidebook, Greater Green River Basin Symposium (E.M. Schell, ed.). p.153-156.

2. Dana, G. F. and Smith, J. W., 1973, Artesian Aquifer, New Fork Tongne of the Wasatch Formation, Northern Green River Basin, 25th Field Conference—1973, Wyoming Geological Association Guidebook, the Geology and Mineral Resources of the Greater Green River Basin, p.201-206.

3. Dana, G. F. and Smith, J. W., 1976, Nature of Black Water Occurrence, Northern Green River Basin, Wy. Geol. Assoc. Earth Sci. Ball., V.9, No.1, p. 9-16.

4. Smith, E. B. Branthaver, J. F. and Heppner, R.A., 1980, Porphyrins in Black Trona Water, Org. Geochem., 2, p.141-152.

5. Yen, T. F., Wen, S. S., Tang, J. I. S., Kwan, J. T., Young,D. K, and Chow, E., 1978, Structural Characterization Of Bitumen and Kerogen for Devonianian Shale, Proc. 1st Gas-Shale Symp. (NTIC), p.572-588.

6. Miknis, F. P., Netzel, D. A., Smith, J. W., Mast, M. A., and Maciel, G. E., 1982, ^{13}C-NMR Measurements of the Genetic Potentials of Oil Shales, Geochim. Cosmochim. Acta, 46, p.977-984.

7. Branthaver, J. F. and Barden, R. E., 1983, ^{13}C-NMR, i.r. and Flurescene Spectroscopic Studies of the Polymeric Acids Found in Black Trona Water from the Green River Basin, Org. Geochem., 4, p.47-120.

8. Duke, R. B. and Seppi, N. F., 1977, Characterization of Wyoming Black Trona Brine, Am. Chem. Soc., Div. Pet. Chem. Prepr., 22(2), p.777-784.

9. Fester, J. I. and Robinson, W. E., 1966, Oxygen Functional Groups in Green River Oil Shale Kerogen and Trona Acids, Adv. Chem. Series, 55, p.22-31.

10. Tarden, R. E., Logan, E. R., Branthaver, J. F. and Neet, K. E., 1984, The Average Molecular Weight and Shape of the Polymetric Acids Found in Black Trona Water from the Green River Basin, Org. Geochem., 5, p.217-225.

11.   Gallegos, E. J., 1973, Identification of Phenylcyclo-
       paraffin Alkanes and Other Monoaromatics in Green
       River Shale by Gas Chromatography–Mass Spectrometry,
       Anal. Chem., 45, p.1399–1403.

12.   Anders, D. E., Doolittle, F. G. and Robinson, W. E.,
       1973, Analysis of Some Aromatic Hydrocarbons in a
       Benzene Soluble Bitumen from Green River Shale, Geo-
       chim. Cosmochim. Acta, 37, p.1213.

13.   Philp, R. P., 1985, Fossil Fuel Biomarkers, Applica-
       tions and Spectra, Elsevier Sci. Publishers, Amster-
       dam, p.294.

14.   Mackenzie, A. S., Hoffmann, C. F. and Maxwell, J. R.,
       1981, Molecular Parameters of Maturation in Toarcin
       Shales, Paris Basin, France, III – Changes in Aro-
       matic Steroid Hydrocarbons, Geochim. Cosmochim. Acta,
       45, p.1345–1355.

15.   Tissot, B. P. and Welte, D. H., 1984, Petroleum For-
       mation and Occurance, 2nd ed., Springer-Verlag, Ber-
       lin, p.105.

16.   Brassell, S. C., Eglinton, G., Maxwell, J. R. and
       Philp, R. P., 1978, Natural Background of Alkanes in
       the Aquatic Environment in Aquatic Pollutants,
       (Hutzinger, O., Van Lelyveld and Zoeteman, ed.)
       Pergamon Press, Oxford, p.69–86.

17.   Welte, D. H. and Waples, D., 1973, Uber die Beror-
       zugung Geradzahliger n-Alkane im Sedimentgesteinen
       Naturwiss, 60, p.516–517.

18.   Smith, J. W., 1974, Geochemistry of Oil Shale Genesis
       in Colorado Piceance Creek Basin, Rocky Mountain
       Association of Geologists, 1974 Guide Book, p.71–79.

19.   Henderson, W., Wollrab, V. and Eglinton, G., 1968,
       Identification of Steroids and Triterpanes from a
       Geological Source by Capillary Gas-Liquid Chromatog-
       raphy and Mass Spectrometry, Chem. Comm., p.710.

20.   Ourisson, G., Albrecht, P. and Rohmer, M., 1984, The
       Microbial Origin of Fossil Fuels, Sci. American, 251,
       p.44–51.

21.   Brassell, S. C. and Eglinton, G., 1983, Steroids and
       Triterpenoids in Deep Sea Sediments as Environmental
       and Diagenetic Indicators, Adv. Org. Geochem., p.684-
       697.

22.   Gallegos, E. J., 1976, Biological Fossil Hydrocarbons
       in Shales, in Oil Shale, (Yen, T. F. and Chilingarian,
       G. V., ed.) Elsevier Sci. Publishers, Amsterdam, p.
       149-180.

23.   Mackenzie, A. S., 1984, Applications of Biological
       Markers in Petroleum Geochemistry in Adv. Petroleum
       Geochemistry, Vol. 1, (Brooks, J. and Welte, D. ed.)
       Academic Press, London, p.115-214.

24.   Nes, W. R., 1974, Role of Sterols in Membranes,
       Lipids, 9, p.596-612.

25.   Moldowan, J. M., Seifert, W. K. and Gallegos, E. J.,
       1983, Identification of an Extended Series of Tricy-
       clic Terpanes in Petroleum, Geochim. Cosmochim.
       Acta, 47, p.1531-1534.

26.   Anders, D. E. and Robinson, W. E., 1971, Cycloalkane
       Constituents of the Bitumen from Green River Shale,
       Geochim. Cosmochim. Acta, 35, p.661.

27.   Gallegos, E. J., 1971, Identification of New Steranes,
       Terpanes, and Branded Paraffins in Green River Shale
       by Combined Capillary Gas Chromatography and Mass
       Spectrometry, Anal. Chem., 43, p.1151.

28.   Reed, W. E., 1977, Molecular Compositions of Weathered
       Petroleum and Comparison with its Possible Source,
       Geochim. Cosmochim. Acta, 41, p.237-247.

29.   Aquino Neto, F. R., Restle, A., Albrecht, J. and
       Ourisson, G., 1982, Novel Tricyclic Terpanes (C_{19},
       C_{20}) in Sediments and Petroleum, Tet. Letts., 23,
       p.2027-2030.

30.   Ekweozor, C. M. and Strausz, O. P., 1982, 18, 19-
       Bisnor-13 H, 14 H-Cheilanthane, A Novel Degraded
       Tricyclic Sesterterpenoid-Type Hydrocarbon from
       Athabasca Oil Sands, Tet. Letts., 23, p.2711-2714.

31. Trendel, J.-M., Restle, A., Connan, J. and Albrecht, Pł, 1982, Identification of a Novel Series of Tetra-cyclic Terpene Hydrocarbons (C_{24}-C_{27}) in Sediments and Petroleum, Chem. Comm., p.304-306.

32. McDonale, R. E., 1972, Eocene and Paleocene Rocks of the Southern and Central Basins in Geologic Atlas of the Rocky Mountain Regions, (Kellog, W. W. ed.) p.243-256.

33. Bendoraitis, T. G., 1973, Hydrocarbons of Biogenic Origin in Petroleum -- Aromatic Triterpenes and Bicy-clic Sesquiterpenes, 6th International Meeting on Organic Geochemistry, Rueil-Malmaison.

34. Cyr, T. D. and Strausz, O. P., 1983, The Structures of Tricyclic Terpenoid Carboxylic Acids and Their Parent Alkanes in the Alberta Oil Sands, Chem. Comm., p.1028 -1030.

INDEX